Gender–Technology Relations

Also by Hilde G. Corneliussen

DIGITAL CULTURE, PLAY, AND IDENTITY: A World of Warcraft® Reader
(*co-edited with Jill Walker Rettberg*)

Gender–Technology Relations

Exploring Stability and Change

Hilde G. Corneliussen
Associate Professor, University of Bergen, Norway

First published 2012 by
PALGRAVE MACMILLAN

Palgrave Macmillan in the UK is an imprint of Macmillan Publishers Limited, registered in England, company number 785998, of Houndmills, Basingstoke, Hampshire RG21 6XS.

Palgrave Macmillan in the US is a division of St Martin's Press LLC, 175 Fifth Avenue, New York, NY 10010.

Palgrave Macmillan is the global academic imprint of the above companies and has companies and representatives throughout the world.

Palgrave® and Macmillan® are registered trademarks in the United States, the United Kingdom, Europe and other countries.

ISBN: 978–0–230–30013–2

This book is printed on paper suitable for recycling and made from fully managed and sustained forest sources. Logging, pulping and manufacturing processes are expected to conform to the environmental regulations of the country of origin.

A catalogue record for this book is available from the British Library.

A catalog record for this book is available from the Library of Congress.

10 9 8 7 6 5 4 3 2 1
21 20 19 18 17 16 15 14 13 12

Contents

Acknowledgements

Earlier versions of Chapter 2 appeared in Sveningsson Elm, M. and Sundén, J. (Eds) (2007) *Cyberfeminism in Northern Lights* (Newcastle upon Tyne: Cambridge Scholars Publishing) and in Misa, T.J. (Ed.) (2010) *Gender Codes: Why Women Are Leaving Computing* (Hoboken, NJ: IEEE Computer Society and John Wiley & Sons, Inc.), and has been published with the permission of Cambridge Scholars Publishing and IEEE Computer Society.

An earlier version of Chapter 3 appeared in *Kvinneforskning* 27 (3) 2003 and has been published with the permission of Kilden.

Part of Chapter 4 appeared in Lie, M. (Ed.) (2003) *He, She and IT Revisited. New Perspectives on Gender in the Information Society* (Oslo: Gyldendal Akademisk) and has been published with the permission of Gyldendal Akademisk.

Earlier versions of Chapter 5 appeared in Archibald, J., Emms, J., Grundy, F., Payne, J. and Turner, E. (Eds) (2005) *The Gender Politics of ICT* (Middlesex: Middlesex University Press) and in *Journal of Information, Communication and Ethics in Society* 3 (4, Special Issue: Women in Computing, 2005) and has been published with the permission of Middlesex University Press and Emerald Group Publishing.

Writing this book has been a long process, and there are many people to thank for their involvement, help and inspiration, from colleagues at my home department, Humanistic Informatics/Digital Culture, to the international network of Gender and ICTs researchers that has met regularly at conferences and provided an important arena for testing ideas and theories. A special thanks to Christina Mörtberg for reading and commenting on Chapter 1, Thomas J. Misa for his helpful comments and proofreading of an earlier version of Chapter 2, Andy for proofreading Chapter 6, and the two anonymous reviewers from Palgrave Macmillan, as well as anonymous readers for earlier versions of Chapters 2–5. I also want to thank Felicity Plester and Catherine Mitchell at Palgrave Macmillan for their help and support in the process of finishing this book. Finally, my husband Atle C. Gandrudbakken – always there, always supportive.

1
Disrupting the Impression of Stability in Gender–Technology Relations

Introduction

Information and communication technologies (ICT) have become vital in today's society to such an extent that they even figure as a key symbol 'defining' our time (Lie, 1998). Technology is, however, not simply artefacts or technological systems operating according to technological logics, but is also part of what we will explore here as gender-technology relations. Men's and women's relationships to technology have been marked by difference; in fact, we have been talking about the 'digital gender gap' in the Western world since the late 1970s (Cockburn, 1983; Mörtberg, 1994; Wajcman, 2004). We have seen research trying to reveal the mechanisms of this gender gap (Hacker, 1989; Lie and Sørensen, 1996b; Håpnes and Rasmussen, 2003; Cohoon and Aspray, 2006b), discussions on how to 'bridge' it (Sørensen, 1992; Salminen-Karlsson, 1999; Margolis and Fisher, 2002), a number of initiatives to encourage girls and women to become more active computer users (Stuedahl, 1997a; Lagesen, 2003; Corneliussen, 2003b; Faulkner and Lie, 2007), and efforts to recruit, include and retain women in computer education and occupations (Gansmo et al., 2003a; Cohoon, 2006; Trauth et al., 2009). We have seen the digital gender gap being redefined from 'a problem with girls and women' to a problem with institutions and cultures of technology (Turkle, 1988; Håpnes, 1997; Gansmo, 2004, p. 96; Sagebiel and Dahmen, 2006; Bagilhole et al., 2008), and we have seen recent reports stating that the gap is about to disappear in certain fields (Gansmo et al., 2003a; Trauth et al., 2010). Obviously, a great deal has changed regarding ICT since the late 1970s, but despite

1

the changes, we still worry about the digital gender gap – a worry that is often followed by a 'stability argument', pointing to how the situation has not changed.

'The cultural association between masculinity and technology in Western societies is hard to exaggerate. It operates not only as a popular assumption...but also as an academic "truth"', Keith Grint and Rosalind Gill claimed in 1995 (p. 3). When we review research literature on women in engineering and technology, in particular in computer-related disciplines and professions, it still seems to hold a degree of truth, as a recurring argument in this literature is either that the situation for women has 'still' not changed despite continuous efforts since 1980 to improve the situation (Ahuja, 2002; Adam et al., 2005; Webster, 2005; Sagebiel and Dahmen, 2006; Bagilhole et al., 2008) or that temporary progress 'has stalled or eroded' (Wasburn and Miller, 2006). Mainstream technology researchers and theorists, such as Jacques Ellul (1964), Nicholas Negroponte (1995), Manuel Castells (1996) or Paul Virilio (2000), emphasize the huge changes brought to modern society by computer technology – some with optimism, others with less optimism or outright pessimism, not unlike the way in which earlier technological inventions were met with mixed responses (Nye, 2004). What they share is a notion of change, as Charlie Gere points out: 'One of the concomitants of our current digital culture is the sense of rapid change. ... The only thing that never changes is that everything always changes' (2008, p. 7). In contrast to this focus on change we find an emphasis on stability among feminist technology researchers – not in modern society, but in gender-technology relations, as pointed out by Grint and Gill: 'In technology theory the key question has been how to explain change, while for feminists it seems more urgent to explain continuity, the enduring inequalities and the fact that gender relations survive so little changed through every successive wave of technological innovation' (1995, p. 21). The claim of stability has continued until today, both in Europe and the US, and has continued despite changes in attitude to both gender and ICTs. In the US, J. McGrath Cohoon and William Aspray point to how 'almost thirty years of efforts have failed to produce a sustained increase in women's participation in computing. Women remain seriously underrepresented, and the intentions of college-bound students...indicate that the situation is not likely to improve any time soon' (2006a, p. 139). In Europe, Sally Wyatt points out that 'ICTs themselves are different from what they were 20 or 30 years ago, gender relations have changed somewhat, and our theoretical understanding of gender and ICTs has also changed. And yet gender inequalities persist, even in countries such as Norway, and technologies remain implicated in the structure and performance of inequality' (2008a). In Norway, a researcher commenting on her own research showing gender differences in computer use among Norwegian youth says, 'I think that it [the gender differences] will always be like that'.[1] I have also myself emphasized stability, for instance by pointing out that 'discursive constructions of gender and ICT are remarkable stable' (Corneliussen, 2003b). The 'stability argument' is mostly used in relation to ICT education, but it is also applied to other aspects and contexts of ICTs. It refers to a gender gap in numbers, in activities and in discourses. As the quotes illustrate, the gender differences even appear to be something that is not likely to change – not soon, or not ever.

The emphasis on stability is not a general trend within feminist technology research, and a number of major research projects during the 1990s and early 2000s have documented change in gender-technology relations, as well as variation and non-hegemonic expressions. In a Nordic context, Christina Mörtberg, studying women in information technology, has shown how women are 'transgressing boundaries', illustrating that the 'male dominance is cracked in places' and that women are not only being shaped by, but also contribute to the shaping of information technology (1997). Merete Lie illustrates in *Computer Dialogues. Technology, Gender and Change* how changes from mechanical machinery to information technology have challenged and changed images of gender and technology (1998). In this work, she found that information technology can be supportive, or not, for both males and females. Tove Håpnes and Bente Rasmussen illustrated young Norwegian girls' profound interest and pleasure in the 'new' Internet technologies in the mid-1990s, illustrating changes in girls' relationship to technology (1999, 2003). In the early 2000s, Helen Jøsok Gansmo et al. (2003a) found that the gender gap in terms of access to and use of ICTs was already melting away among the younger age groups in Norway, and Hege Nordli, who searched for the female hackers, found young women with an 'intimate and close relationship' to the computer (2003). Pirjo Elovaara illustrates change both in technological equipment and in user relations, starting her story with 'mere terminals' and ending up with ICTs where those that used to be called 'users' have become 'developers and designers of information technology' (2004, p. 13). Feminist technology researchers have also asked for change within the field, and even aimed at creating it (Bratteteig and Verne, 1985; Björkman, 2002; Schiebinger, 2008a; Hayes, 2010b). Thus, despite a number of research projects documenting changing gender-technology relations, it is striking, if not a paradox, that the claim about stability continues to re-emerge.

My recent frustration over the 'stability argument' has grown out of a dual experience: my long-lasting interest in reading literature on gender and ICTs, and a more recent engagement in an EU project reviewing literature on gender and science between 1980 and 2007.[2] The 'stability argument' as it appears in literature on gender and ICTs is often presented either in the introduction (as a justification for continuous studies in this field) or in the conclusion (with a deep sigh over persistent inequalities). In both cases it makes depressing reading, with observations of how efforts to improve the situation have failed, still leaving women as a minority in important fields of technology. The EU project, though focusing more broadly on science, clearly illustrated a continuous heartfelt need to improve women's situation in research, including the field of technology, since 1980. However, the situation is not the same today as it was in 1980 (Egeland, 1999). The extent of this change is illustrated in an article in the Norwegian journal for women's research *Nytt om kvinneforskning* from 1994, where the author reviewed 10 years of struggle to achieve gender equality in academia. To illustrate the many creative forms this struggle has taken, she quoted a letter sent by a female research recruit to the Norwegian minister of education in 1984. The letter is short and concise, declaring that there has been a restructuring of the University of Trondheim, followed by an election for the new collegium. The author lists the 17 newly elected members of the board, all in an informal tone, giving only their first names: they were all men, as was the minister of education. The letter ends by simply stating 'That was all for this time, compliments Agnes' (Pedersen, 1994). The only female name belongs to the sender, pointing out something so obvious that neither comment nor argument is required. Agnes' reason for writing to the minister was not that she thought he would wish to know the first names of the board members: by stripping away all other information, gender alone is highlighted, making visible a striking lack of women in positions of power, control and status at the university. Today there are regulations to ensure that both genders are represented on Norwegian university boards (Holter, 2010), as well as in public organizations and private corporations. Even though women are still in a minority in higher academic positions in Norway, as in most Western countries (European Commission, 2009), a number of circumstances have changed. These range from rules and regulations to perceptions of justice/inequality, evaluation of measures to deal with inequality, and a changed reality: there are more women on boards and in academic positions than there were in 1984 (Corneliussen, 2009). Changing attitudes among authorities are also documented in an increasing number of funded research projects that focus on women's situation in academia in general, and in technology and computing in particular (Bagilhole et al., 2008). Changes thus cover several layers, from attitudes and written regulations to actual outcomes.

I do not propose that standards be lowered or that the goal of feminism be reduced from gender equality to women simply being a little less compromised; neither do I claim that the stability argument is false. No matter how much we analyse this, it will still be true that there are few women in certain technological fields, such as computer science. My aim is rather to explore how research might be presented in ways that allow certain meanings to dominate, and to explore how research produces certain stories (White, 2003). My own interest originates from my first encounter with the topic in 1997 and lies in how such results are produced, and how meaning is constructed. More particularly, my focus is on how gender-technology relations are perceived, described and thus shaped in certain ways. I was asked to comment on a report on 'Computers and Women' by the Norwegian Ministry of Children and Family Affairs, and met with this description of the situation: 'During the last 20 years, Man has been moved to take more interest in children and the home – Woman now has to take more interest in technology and industry. Technology and industry are primary elements in society, with very old ancestors; they offer manifold and interesting challenges; and competitive industry is in fact the foundation for a high standard of living' (Reland, 1997, p. 16). This quote, filled with contempt for women's passivity in relation to technology, made me feel angry, disappointed and ashamed. Angry over the ignorance of women's historical contributions to industrialization in the Western world, with their hard work as cheap labour; disappointed by the lack of respect for women's contributions to family life, thus also enabling men to work outside the home and, consequently, to dominate public positions of status and power; and ashamed at this coming from a government office as a report about the situation of women and computer technology in Norway in 1997. What puzzled me most was not the reference to men as interested in technology and women as uninterested, but the description, the words used to construct one particular version of this story. The author *could* have made other choices in his description of women; he could have mentioned that women were not allowed to enter higher education until the late 1880s, that they have been excluded from taking part in many important fields where technology has been developed and used.

Or that Norway had, and still has, a highly gender-divided workforce, with women dominating in service, health and caring occupations, and in low-paid and part-time positions.[3] Or perhaps that housework and child care – despite men being 'moved' – is still unequally distributed between men and women (Holter et al., 2009). Pondering on why the author chose words filled with contempt and disregard for Norwegian women's contributions, as if he wanted to make women feel guilty for not having contributed to their country's development, is what set me off exploring computing history in greater detail. It was not the actual activities, the actors, or the development of computer technology that interested me; rather, it was the stories about the appropriation and diffusion of technology, and the construction of meaning, of cultural significance, that took place in those stories. And that is what this book is about; motivated by the frustration over the minister's ahistorical description of the situation, I have spent more than a decade exploring the construction of meaning related to the cultural appropriation of computer technology in Norway since the early 1980s.

Social sciences do not represent a neutral monitor of society, but focus on particular topics, events, groups and social factors, and this is done with specific research lenses, producing specific results and consequences. It might be true that there are few women in computer science, but are there other ways of seeing the situation? What is it about gender and technology in general, or gender and ICTs in particular, that makes it natural to talk about lack of change, or even to predict a never-ending difference? And how can gender-technology relations remain unchanged when everything around and within is changing? This book will question the impression of stability, not by rejecting research that has documented stability in gender-technology relations, but as a thought experiment, to explore what a specific focus on change can teach us. Will it show that gender-technology relations have indeed remained stable since the 1980s, despite severe changes in technology, its diffusion and use? Or could it be that feminist literature has developed some internal truths, as Grint and Gill (1995) propose?

Theories predicting change

Theories inspired by post-structuralism have described both gender and technology as multi-layered categories. Sandra Harding's proposal to view gender as having three aspects, gender symbolism, gender structure and individual gender (1986), has been widely referred to as a useful starting point for analysis of gender-technology relations. Similarly, recent technology research has emphasized that technology should not only be seen as the artefact itself, but as a complex that includes knowledge, routine and symbols in a 'seamless web' of society and technology, and with considerable 'interpretative flexibility' (Hughes, 1986; Bijker and Law, 1992; Pinch and Bijker, 1997 [1987]). Furthermore, gender and technology researchers have emphasized that gender and technology are co-constructed, woven together in an intricate web of society, technology, culture and nature (Lie and Sørensen, 1996a; Faulkner, 2001; Lie, 2003a; Faulkner and Lie, 2007). This analytical framework, Wendy Faulkner claims, indicates that 'one cannot understand technology without reference to gender, so one cannot understand gender without reference to technology' (2001, p. 90). In other words, we have seen an increasingly complex understanding of how gender-technology relations work. And in the very heart of post-structuralist theories lies the potential for change; the de-construction of social and cultural meaning that has been constructed in certain ways, based on choices, as effects of actions and systems, not based on a fixed essentialism of gender. Thus, post-structuralist theories are pointing to how meaning is constantly constructed and reconstructed, and that matters could have been constructed in other ways (Søndergaard, 2002), or that 'technological artifacts could be other' (Faulkner, 2001, p. 83). The recent dominant notions of gender as fluid and flexible (Haavind, 1994; Søndergaard, 1996; Braidotti, 2002), such as Bob Connell's description of gender as something that is constituted in a 'historical process, and accordingly can never be fixed, nor exactly reproduced' (2002), also seem to reject the idea of stability from the start. Thus, when it comes to gender and ICTs, there seems to be a gap between the theoretical and the empirical level; the first approach theorizes and predicts a constant change, while empirical research gives nourishment to descriptions of stability.

For decades feminists have criticized mainstream technology research, political authorities and educational institutions for being gender-blind. To substantiate the need for a revised history of technology, Autumn Stanley points out how mainstream technology researchers, when 'looking backward through the distorted glass of a prevailing cultural stereotype that women do not invent, have found, not surprisingly, that women never did invent' (1998 [1983]). On the other hand, Connell has pointed out how even 'sex difference research' is blind to seeing similarities (2002). My question here is whether we have become blind to seeing change in gender-technology relations: could it be that we have been 'looking for' stability instead of change, or for gender differences instead of similarities or 'differences within'? Is it time to scrutinize how we, through our research, retell, reconstruct and thereby also re-create a specific version of 'reality'? Or as Wyatt puts it: '[t]he question is not so much "do we need further analysis?" but rather, "what kind of research do we need?" ' (2008a)

We do of course need research of different kinds, and the aim here is to ask what we can learn by focusing on change in the relation between gender and technology, or more specifically, gender and ICTs. We certainly know that gender-technology relations have not remained stable. This is obvious when we take on a long-term historical perspective, but also since the 1980s, even since the turn of the millennium, there have been significant changes in gender-technology relations. What kind of changes can be seen, on what levels, in which fields, and under which conditions? Recent gender and ICTs research has pointed out the need to recognize diversity, variations, and multiple masculinities and femininities. But are these variations change, or are they seen as examples of exceptions, special cases, or extraordinary individuals? Does change need to involve a majority, a consensus, or perhaps 'relevant social groups'? Does it need to be permanent, include a 'critical mass', or go in a particular direction to be recognized as change?

In this book we will explore questions of change and stability on several levels. We will look at change over time in cultural images and popular discourse, and we will explore how perceptions of gender and technology have changed within higher ICT education. Men and women, masculinity and femininity will be explored, and we will see difference not only between men and women, but also between men and between women, as well as change in individuals' relationships to computer technology.

Inequalities persist, even in Norway

The empirical studies that provide the foundation for the discussions in the forthcoming chapters are all from Norway, a country that offers an interesting site for the study of gender-technology relations. Norway is a rich country that, along with the other Nordic countries in the far North, ranks high on global gender equality measures and statistics (Holter, 2010), as with the Global Gender Index, ranking countries according to men's and women's access to resources (Hausmann et al., 2010).[4] There is a political drive towards gender equality, and 'state feminism' (Hernes, 1987) (with a strong mainstreaming of gender perspectives in politics, white papers and educational institutions) is often mentioned as one of the main driving forces towards gender equality in Norway (Gansmo et al., 2003b). Thus, we should perhaps expect that a country such as Norway also stands out as exceptional when it comes to gender and technology or ICTs. However, the situation of women in technology and computing is not equally impressive. Various similarities between Norway and other Western countries have been pointed out regarding the situation of gender and ICTs, such as the low proportion of women in computing (Charles and Bradley, 2006; Cohoon and Aspray, 2006b); the computer as a boy's toy (Nissen, 1996; Mahoney, 2001; Mellström, 2004); the symbolic importance of masculine stereotypes, such as the male hacker (Lagesen, 2003; Hayes, 2010b); and stereotypes associating men with technology (Phipps, 2007).

The fact that Norway shows similarities with other Western countries, in particular concerning a low proportion of women in the field of technology, has made it a good example for the 'stability argument', as Wyatt illustrates in pointing to how 'gender inequalities persist, even in countries such as Norway' (2008a: cf. Webster, 1996). As we will see, gender equality is not only about equal rights, policy and regulation: a growing number of international studies have documented that gender equality does not always follow expected patterns. Reports from the EU project *Meta-Analysis of Gender and Science Research*[5] found the level of gender occupational segregation[6] in Norway to be 2 per cent higher than the EU-average of 25.2 per cent, and higher than in the UK (25.3 per cent) and Romania (23.3 per cent) (Meulders et al., 2010). Romania was, however, ranked number 67 on the 2010 Global Gender Index (GGI) and the UK number 15 (Hausmann et al., 2010). While the EU-27 average percentage of female researchers is 30, Norway achieves slightly above this figure with 33 per cent, but well below Lithuania (49 per cent), Latvia (47 per cent) and Bulgaria (45 per cent) (European Commission, 2009). These three countries are, however, ranked, respectively, as numbers 35, 18 and 50 on the GGI for 2010. At the other end of the scale, we find the Netherlands and Luxembourg with barely 18 per cent of researchers being female (numbers 17 and 26 on the GGI). The EU-average for the proportion of female PhD graduates in science, mathematics and computing (European Commission, 2009) is 41 per cent, but only 31 per cent in Norway – far below Italy (52 per cent), UK (38 per cent) and France (37 per cent). Italy was ranked as number 74 and France as 46 on the GGI (Hausmann et al., 2010).

There is, however, one field in which the Nordic countries score high in comparison with the EU-27: the proportion of women sitting on boards. For Norway, this is due to regulations requiring that at least

40 per cent of board members are women, and Norway ranks second with women comprising 45 per cent of board members in higher education, behind Sweden (49 per cent), and well above the UK (25 per cent) and the EU-27 average (22 per cent). Norway is also ranked top (32 per cent) with regard to women as heads of institutions in the higher educational sector, well above the EU-27 average (13 per cent) and Denmark (barely 5 per cent) (European Commission, 2009, pp. 97, 99).

A feminization of the student population has been found throughout the Western world, with a higher number of female students in most European countries (European Commission, 2009). Advancing through the ranks, to PhD students and the higher ranks of academic staff, however, the proportion of women decreases for each level, with only 18 per cent of women in grade A academic positions (equivalent to a full professorship in most countries), illustrating the 'leaky pipeline', 'glass ceiling' and 'sticky floor'. The Glass Ceiling Index (GCI) for European countries presented in *She Figures 2009*, indicating women's access to grade A positions, shows that the GCI remained stable or improved for women in all countries but Norway between 2004 and 2007 (European Commission, 2009, p. 68).

The statistical examples presented above illustrate a complex picture, and explanations for the inconsistencies we find when comparing gender statistics cross-nationally have been suggested by, for instance, Øystein Gullvåg Holter (2010). Based on findings of an EU study on working life, Holter concludes that 'organisations in countries with weak welfare and work regulation systems often work more actively to promote gender equality than organisations in more gender-equal countries such as Norway, which tend to be more passive. This is due to a belief that gender equality has already been realized; there is no need to make an active effort, and in any case, the state will take care of it' (2010). Maria Charles and Karen Bradley (2006, p. 194), who found a 'striking cross-national uniformity in the sex typing of computer science programs' in 21 industrial countries, point to how a high degree of gender equality also includes the freedom to choose traditional educational paths (p. 196). Anne-Sophie Godfroy-Genin also found that 'high gender awareness is not always related to a better situation for women' (2009, p. 85). Thus, while a relatively high gender awareness in the Nordic countries has not resulted in high numbers of women in technology, the patterns found in Eastern Europe can be seen in light of how the 'former Soviet ideology officially promoted gender equality but no gender awareness', Godfroy-Genin suggests (ibid.). Godfroy-Genin concludes from the PROMETEA project, focusing on gender in engineering and technology research, that 'similarities among European women researchers in Europe, despite the huge variety of settings, have been the surprising results of PROMETEA' (2009, p. 88). The pattern of a low and decreasing proportion of women in technology- and computer-related fields is an international phenomenon cutting across other patterns and measures of gender equality (European Commission, 2009); these tendencies are found across Europe as well as in the US (Adam et al., 2005; Webster, 2005; Trauth and Howcroft, 2006; British Computer Society (BCS) e-skills UK and Intellect, 2009; Ashcraft and Blithe, 2010; Hayes, 2010a). The international trend of women's under-representation in technology and computer-related fields is reflected in more than numbers. There are also cross-cultural similarities in gender stereotypes and the effects they have on girls' and women's relationships to computer technology (Phipps, 2007). As already noted, there are similarities in how the continuously low proportion of women generates a rhetoric of stability.

It is necessary to explore local, national and regional situations regarding gender and technology, as pointed out by Eileen Trauth et al. (2009). However, the cross-national patterns of women's situation in technology and computing also make this a field where we can benefit from comparing the situation in different countries. Thus, while the main empirical material in the forthcoming chapters is drawn from Norway, the questions asked, and the discussions and findings should be of interest to an international audience. The literature references throughout the book point to similar findings and discussions in other countries.

Research lenses: focus and blind spots

Research is a process that starts with assumptions and prior knowledge (Popper, 1974; Haavind, 2000). Most often, particular theories and methods are also applied, which further specifies our understanding of the field we intend to study. These assumptions, previous knowledge and theories help us decide what or who it is interesting to study. Thus, different starting points and theories steer our research focus towards certain contexts that appear to be worthy of study, along with certain elements or factors that appear to be of importance to the analysis. Alternative contexts, elements or factors are omitted as having lesser importance and remain in the researcher's blind spot. In studies of gender and technology, we need theories about how the relationship between gender and technology should be perceived, how gender and technology relate to each other and how they affect each other. As we will see, different feminist theories have created different interpretations of gender-technology relations.[7]

Most often women's presence is what has triggered an interest in using gender as an analytical category, while male-only contexts have often been perceived as ungendered contexts. While women have a tendency to appear as gendered, men often appear to illustrate an ungendered norm, in research as in politics (Scott, 1996; Lohan, 2000). Women and technology have not always been 'found' in the same environment. Thus, mainstream technology research focusing on technological contexts where women have been a minority has rarely perceived gender as an important factor, while early feminist research focusing on women's situation rarely found technology important. This also relies on a specific understanding of what kind of technology has been found relevant to study (Faulkner, 2001, p. 85). The cultural and symbolic connection between men and masculinity and technology has made technology appear as 'what men do and women do not' (Oldenziel, 2001). Since the 1980s feminist technology research has grown significantly, showing, for instance, that women also have strong relationships to technology, given that we accept a broader definition of technology that also includes technology used by women (Wajcman, 1991, p. 137), such as household technology or technology affecting women's lives, such as reproductive technology. Anne-Jorunn Berg (1996, 1999) has, for instance, documented how women can have very intimate relationships with technology they use in the home, illustrated by a woman who gave her tumble dryer a name and talked affectionately about it as though it was a friend. Autumn Stanley even claims that many of today's technologies were originally invented by women, but were not considered 'technology' before men took over the power and control. What we need is a revised history of technology based on a new definition of technology, changed from 'what men do to what *people* do', Stanley claims, which would also change our perception of 'significant technology' to include technology used by women (1998 [1983], p. 17).

The alienation between women and technology has made different feminisms propose different ways of understanding and interpreting gender-technology relations. The association between masculinity and technology that Grint and Gill (1995) refer to as an 'academic truth' also applies to most feminist contributions within science and technology studies, which, rather than challenging the close relationship between masculinity and technology, accept it and seek to explain why it is or has come to be that way. Eco-feminism springs from an idea of women as closer to nature than men, and technology produced by men is assumed to be founded on – and reflect – men's domination over both nature and women (Mack, 2001, p. 152). The main critique from eco-feminism has targeted destructive military technology: that which destroys lives as opposed to women's biological ability to create lives (Cox, 1992). Eco-feminism is based on two sets of determinism. First, by ascribing some essential qualities, values and morals to women based on biology, it is treating women as one homogenous group. Out of harmony with feminists' struggle for equality, 'the values ascribed to women by eco-feminists originate in women's subordination; precisely those characteristics which other feminists have tried to historicize and have shown to be contingent – the product of oppression – are here valorized as essentially feminine', Grint and Gill point out (1995, p. 5). Second, eco-feminism involves a deterministic perception of technology, assuming that the internal essence of technology is patriarchal, simply reflecting men's dominance and attempt to control women. When technology is perceived as inherently patriarchal, there is no other option for the eco-feminist than an 'absolute rejection of technology' and the production of a new technology based on women's values (Grint and Gill, 1995). The same emphasis on technology as a tool for patriarchal control over women can also be seen in early studies of computer technology in office automation. While mainstream technology researchers explored how new information technology would have negative effects for male workers, feminist researchers pointed out that it would have other consequences for women than for men. This strand of research has often been labelled 'impact' because of the emphasis on women as more or less passive 'victims' of the new information technology. The fear of women being made redundant by technology was discussed, along with other 'victimizing' effects. Janet Barker and Hazel Downing, for instance, saw a 'transformation of patriarchal relations of control' with the introduction of word processing in office work in Britain in the early 1980s (1985 [1980], p. 161). There were worries about the work itself; a fear that computers would make women's work more routine; there were concerns regarding health issues, in particular related to working in front of the computer screen;[8] there were also concerns regarding gender relations, in particular management, which in principle meant men's control and surveillance of women in work environments (Barker and Downing, 1985 [1980], cf. Webster, 1995).

Another suggestion for how to interpret gender-technology relations has come from liberal feminism, which sees technology as originally gender neutral (Faulkner, 2001). Men and women are seen as equals who by nature share a humanity and rationality. Existing gender differences are explained by reference to stereotypical social gender patterns that act as barriers to women, and gender differences in relation to technology are explained by men's and women's different social positions in relation to technology. Thus, while the eco-feminist argument is based on men and women as fundamentally different, the liberal feminist argument is based on men and women as essentially equal. That also makes the political message from liberal feminism more positive and less prone to reject technology as such; liberal feminists imagine that the situation can change without creating an isolated women's culture or women's technology. Liberal feminism has not been without criticism, for instance, it has been said to have a rigid concept of technology that appears to be unaffected by social categories such as gender. Other critiques have pointed to how the problem is defined as a 'women's problem' due to the idea of technology as neutral and men and women as equal. If everything else is neutral and similar, the solution appears to be for women to change (Grint and Gill, 1995).

Eco-feminist theory of difference and the liberal feminist idea of a neutral technology have both been criticized. Early feminist technology researchers have pointed out that gender-technology relations are constructed through historical and cultural processes that define technology as masculine (Cockburn, 1985; Wajcman, 1991; Grint and Gill, 1995, p. 8; Oldenziel et al., 2003). In this tradition, technology is seen not simply as a physical artefact, but also as skills, knowledge, routines and symbolic images (Wajcman, 1991, p. 149). In this perspective on technology as being socially shaped, technology is seen as a product of culture and society, which calls for a contextualization of gender and technology, as well as taking into account how technology is gendered in different ways. In her famous study of gender constructions in the print industry, Cynthia Cockburn emphasized capitalism's negative impact on women's participation in technological cultures. Women lost out both as women and as workers, because men monopolized technological competence, shutting women out from the opportunity to acquire the necessary technical skills (Cockburn, 1985, p. 39).

Masculinity and technology have become symbolically intertwined, it has been claimed, to such a degree that they both could refer to each other (Grint and Gill, 1995; Lie, 1998; Faulkner, 2001). On the other hand, technology is affected by the 'technological circle', according to Cockburn, where it is both source to, and sign of, women's suppression. The result is a technology that is not neutral, but rather is 'inscribed' with the social relations of its production and use (Cockburn, 1992, p. 38). Thus, technology supports, as part of culture, patriarchal values suppressing women, and therefore also contributes to the maintenance of gender patterns in society (Grint and Gill, 1995, pp. 8, 10). According to Judy Wajcman's early writing, the gender differences in relation to technology are an extension of the male domination that permeates Western culture, and while technology appears to belong within a masculine culture, women have to give up some of their femininity in order to participate (Wajcman, 1991, pp. 19f). Sherry Turkle has also made similar claims with regard to computer cultures; in particular, she observes computer education as a field where women, in order to participate successfully, have to renounce their preferences – to 'fake it' – and instead adhere to what she described as a masculine culture of computing (Turkle, 1988, p. 41). In a study of computer cultures, Sherry Turkle and Seymour Papert claim that 'women are too often faced with the not necessarily conscious choice of putting themselves at odds either with the cultural associations of the technology or with the cultural constructions of being a woman' (1990, p. 14). Thus, seeing technology as a masculine culture also raises questions concerning identity. Both Wajcman and Turkle have shown that masculinity implies a close relationship to technology, including computers, while feminine identity implies a distance from technology, and thereby also a lack of technical competence (Turkle, 1988, p. 41; Wajcman, 1991, p. 155; Wajcman, 2004, p. 112). The circular symbolic value of technology and masculinity, in which they refer to each other, means that men can use technology to signify masculinity (Lie, 2003b, p. 259). Technology does not do anything similar for women, according to Wajcman: 'Women's identity is not enhanced by their use of machines' (1991, p. 89). Thus, gender identity seems to act as one of the mechanisms reproducing gender difference in relation to technology.

Different from the technological determinism in eco-feminism, Wajcman imagines a possibility of changing women's relationship to technology, but this requires deep changes in society, including change in social patterns that lead to women's suppression. Women's situation will improve when men give up 'the privileges and power that go with this construction of masculinity' (Wajcman, 1991, p. 164; 2004, p. 112). Wajcman's political message is not to change women, as liberal feminism proposes, but to produce new technology. The new technology is not to be produced according to the eco-feminist programme, but based on 'socially desirable values' that surpass the differences between masculinity and femininity (Wajcman, 1991, p. 166).

In 1995 Grint and Gill pointed out that most 'feminist research remains within, rather than outside, the ideological problematic set by the assumption that masculinity and technology are intimately related' (1995, p. 4). Today, however, we can also find attempts to see recent developments in ICTs, in particular the Internet, as a tool for women's liberation and empowerment, for instance in cyberfeminism. Sadie Plant, one of the advocates of cyberfeminism, claims that the Internet, as opposed to earlier technology, represents new opportunities for women, because the new self-organizing technological systems mean that men are no longer in control, but become subordinated to the machine. 'Complex systems and virtual worlds are not only important because they open spaces for existing women within an already existing culture, but also because of the extent to which they undermine both the world-view and the material reality of two thousand years of patriarchal control' (Plant, 2000, p. 265). Cyberfeminism shares some of the eco-feminist celebration of feminine values as better than masculine values. Plant even suggests that Turing was able to design the Turing machine as a result of being forced to take the feminizing hormone oestrogen after having been prosecuted for homosexual acts (Plant, 2000, p. 272).[9] The technophoric optimism of cyberfeminism also resembles technology optimism from the early 1970s, as illustrated by Shulamith Firestone in *The Dialectic of Sex: The Case for Feminist Revolution*, where she imagined that new technology would liberate women from both the tyranny of biology (reproductive technology) and their economic dependency on men (information technology) (1970).

In more recent feminist technology research, inspired by post-structuralism and theories of social construction, there has been an increasing focus on how gender and technology are co-constructed (Cockburn and Dilic, 1994a; Faulkner, 2001; Maass et al., 2007). As mentioned, the relationship between women and technology has not always been readily visible, mainly because women were either absent or a minority in the context of what was considered 'proper' technology. During the 1990s we saw increased interest in studying not only what technology did to women, but also what women did with technology. Faulkner refers to this as a shift from studies of women *in* technology (with a focus on the low proportion of women working with technology) to studies of women *and* technology (emphasizing that more women met technology as users rather than as designers) (Faulkner, 2000b). This shift spurred studies of contexts where women and technology did meet. One of these arenas was the home (Aune, 1996; Berg, 1996), but there was also a growing body of research on girls' and women's relationships with computer technology in other settings (Rasmussen and Håpnes, 1991; Adam et al., 1994; Mörtberg, 1994; Edelman, 1997;

Lie, 1998; Salminen-Karlsson, 1999; Lie, 1998; Rasmussen and Håpnes, 1991; Edelman, 1997). A theory belonging to the tradition of seeing technology and society in a mutual construction is the domestication theory, first proposed by Roger Silverstone et al. (1997 [1992]).[10] They emphasize that a household is not an 'empty space' for new technology to occupy, but has a moral economy; values, interests and routines that shape the household and connect it to society. Domestication of technology should be seen as a mutual process in which the household (or the user) adjusts contemporaneously with new technology, as new technology is adjusted to the household. Strathern emphasizes the cultural presuppositions for new technology in relation to everyday life, and claims that technological artefacts can 'become crucial to the daily routines on which the relationships come to be based ... But they are as much pressed into the enactment of already existing social relationships as they stimulate the creation of new ones. They are thus "domesticated" to social as well as cultural ends' (Strathern, 1992, p. viii).

Domestication theory, as well as other theories dealing with technology as being socially constructed (Bijker et al., 1993 [1987]), have emphasized that technology is not a final developed artefact when it leaves the producer, but rather is in constant development through its use (Oldenziel, 2001). Thus, it is the users' choices and actions, as well as the agreement of important social groups on a technological solution, and not the inherent qualities of the product that decide whether a technology is successful or not (Pinch and Bijker, 1997[1987]). In this perspective, technology is not perceived as neutral, but rather as reflecting temporary meaning created through continuous interpretation. Consequently, technology is assumed to be constructed as masculine or feminine through implementation and use, not by inherent essence.

The consumer is free to define her own ways of using technology, Strathern points out, which is a view even more explicitly developed in script theory, as defined by Madeleine Akrich and Bruno Latour (Akrich, 1992; Akrich and Latour, 1992). Technology leaves the producer with a built-in script that is meant to guide the user. However, users do not always follow the script, but rather create their own – their own ways of using technology that were not originally intended by the producers. Thus, technology is constantly developed in negotiation between the original script from technologists and the users' own creative use (Akrich, 1992). This also illustrates how technology is constructed with particular users in mind, and Cynthia Cockburn and Susan Ormrod (1993) have shown how household technology is constructed with a clearly *gendered* user in mind. They have studied the development of the microwave oven, which first entered the market as a technology intended for men who wanted a quick and easy way to make a meal. Thus, the microwave oven started its career as a consumer product in the 'brown goods' section, among stereo systems, televisions and other entertainment equipment, having clear visual similarities to these. Soon, however, the microwave oven was redesigned to fit the family, and it changed its target group to women (envisaged as housewives)[11]. In the process of this it also changed its location in the department store, moving from entertainment technology to 'white goods', along with other kitchen appliances. The design changed accordingly to fit in with other cooking appliances (Cockburn and Ormrod, 1993; Ormrod, 1994). In short, the microwave oven changed according to assumptions about the main user group, from techno-savvy men to women with a lesser level of technical skills and a higher skills level in cooking.

The case of the microwave oven illustrates how technology is gendered by design and through assumptions about a gendered user all the way from the drawing board to the shelf in the store. Technology designers, however, have been a strongly male-dominated group. In her study of smart houses, a home in which technology is meant to simplify or support daily routines, Berg (1999) shows that technology designers do not always apply knowledge about the target group of their product. Instead, the smart house designers produced technological solutions using an 'I-methodology', taking their own ideas, preferences and interests as a starting point. The result was a smart house more suitable for the 'technically-interested man, not unlike the stereotype of the computer hacker', rather than one designed for the daily routines of housework most often performed by women (Berg, 1999, p. 311).

Despite attempts to escape the essentialism inherent in early feminist technology research, more recent contributions have been criticized for, among other things, having a 'hidden' essentialism. Keith Grint and Steve Woolgar suggest 'subjecting the notion that technology has politics built into it (capitalist, patriarchal or whatever) to the same anti-essentialist critique that is currently used to deny the eco-feminist position that existing technology is inherently patriarchal' (1995, p. 49). Post-essentialist theories are one alternative to the hidden essentialism in constructivist approaches. With 'A Cyborg Manifesto', Donna Haraway (1991a) wanted to create a new language for feminist research and politics, illustrating how hybrid entities such as the cyborg[12] – the human–machine entity – transcend the dichotomies into which we have traditionally divided the world (man/woman, culture/nature, private/public and so on) as simultaneously the transcending elements become incorporated in old and familiar categories (Haraway, 1997, pp. 55f). Haraway argues that humans can no longer be perceived as essentially different from other elements, in nature and science. The difference between culture and nature, human and technology is no longer as obvious as we traditionally assumed, and none of them can be reduced to a single category. The cyborg is not either/or, but both/neither (Bell, 2007, p. 107). 'FemaleMan' is one of Haraway's metaphors on the utopian solution to an assumed universal masculine domination, obtained by incorporating biology, informatics and economy, and thus presenting a challenge to the category of nature. Haraway proposes a 'cyborg anthropology', a study of the web of relations between technology, culture and gender, where the actors are both humans and non-humans (Haraway, 1997, p. 52), thus challenging the traditional epistemological categories that have repeatedly led research into essentialism.

Some critics of post-essentialism claim that it is 'simply beyond the pale' and that the extreme relativism makes it impossible to establish theories about anything (Grint and Woolgar, 1995, p. 65). Gill Kirkup and Laurie Smith Keller claim that the extreme relativism 'suggests that there are as many truths as individual people and that no single truth has any claim to be better than any other' (1992, p. 10, quoted in Grint and Woolgar, 1995, p. 66). 'Feminist objectivity is about limited location and situated knowledge', Haraway claims (1991b, p. 190), pointing out the goal for feminist research as not being relativist as in 'everything goes' but a limited perspective that acknowledges the importance for contextualized and situated knowledge which gives Kirkup and Keller right to claim that there are many truths. However, they are neither accidental nor alternative, but true in a specific local and limited part of reality. To assume that we can see everything in the same gaze is what Haraway calls 'the god-trick', which is not the opposite of her suggested partial vision, but rather an illusion that researchers should seek to avoid (Haraway, 1991b).

Haraway's theories about the cyborg, together with situated knowledge, have inspired feminist technology research (Bell, 2007, pp. 110–12). The cyborg is interesting as a political challenge to a cultural tradition where women's representation more often than not has been coloured by their biological sex, as opposed to men who most often can rest on their social and cultural gender (Scott, 1993, 1996). Haraway's key concepts 'do not represent a "cyborg toolkit"', Daniel Bell points out, but should rather 'be seen as parts of an ongoing working-through of how to talk about "elsewhere"' (2007, p. 128), which points to the hope of 'a more livable place' (Haraway, 2004, p. 1). Haraway's texts have inspired new ways of thinking within feminist technoscience, challenging traditional ways of conceptualizing gender and technology, and gender-technology relations. As pointed out by Cecilia Åsberg and Nina Lykke, feminist technoscience has also been vital in the recent turn towards the physical, referred to under different names, as a material, ontological or postconstructionist turn (2010, pp. 300–1).

This is only a short version of the research history concerning gender and technology, and the point here is not to provide a complete overview, but rather to point out how theories create certain ideas about how relationships between gender and technology work. This makes certain topics interesting to study, certain interpretations plausible, and thereby also makes certain matters visible, while other matters remain in a theoretical blind spot. Gender is rarely visible in research unless it is explicitly defined as an important category, and by not emphasizing gender, descriptions of the world can slip through as 'general' and valid for all human beings. Both political feminism and feminist research have contributed to changes and variations in how to understand and interpret gender-technology relations in comparison with a gender-blind mainstream tradition. This has been achieved by questioning how valid a gender-blind account could be for women, by exploring the close connection between masculinity and technology, and by making women's relationships with technology visible. Feminism has changed disciplines such as science, technology and medicine, as Angela Creager et al. (2001) claim. However, what is considered 'feminist' also changes. Londa Schiebinger points out 'It is a curious phenomenon that when feminist practices or points of view become widely accepted in science and engineering, or in the culture more generally, they are no longer seen as "feminist," but as "just" or simply "true"' (2008b, p. 9).

Feminist research has made a difference compared with mainstream technology research, but feminist research is also itself continually changing in several ways. 'A feminist scientific agenda of one age might be the reactionary agenda of a different age', Scott Gilbert and Karen Rader point out (2001, p. 76, quoted by Creager et al., 2001). The focus of feminist research has changed, what are considered interesting research questions have changed, and theoretical understanding of gender-technology relations has changed. Different feminisms have proposed different interpretations of this relation (Mack, 2001) – all, however, with the shared goal of analysing why women seem to have less strong connections to technology than men. And we have learnt something from all of them, Susanne Maass et al. point out: 'From liberal feminism we have learned to pay attention to women ... From standpoint theory we learned to look for feminine connotated values' and 'hidden norms' contributing to designing ourselves and society, and 'from post-structuralism we have learned the importance of language, of deconstructing the values and dichotomies' (2007, p. 17).

Not only feminist movements, but also feminist researchers might have a political goal, as Wajcman illustrates when concluding her book on feminism and technology, 'intended as a contribution to both academic and political debates about the connections between gender, technology and society' (1991, p. 162). We should not be indifferent to which theories we use, as they also 'can have profound consequences for the practical politics which we adopt' (Grint and Woolgar, 1995, p. 48). Or with Marilyn Strathern's words, quoted by Donna Haraway in an interview: 'It matters which categories you use to think other categories with' (Lykke et al., 2004). Categories are 'thinking technologies that have materiality and effectivity', Haraway explains, and 'are ways of stabilizing meaning in some forms rather than others' (Lykke et al., 2004).

Theories of importance in the forthcoming analysis

The analysis that will be presented in the following chapters owes much to a number of theoretical sources within post-structuralism, some of which have already been mentioned. Post-structuralism is not a coherent theory, but rather represents an idea about how meaning is constructed in society, and there 'are no recipes for creative analyses' in post-structuralism, Dorte Marie Søndergaard reminds us (2002, p. 187). Instead, the researcher has to develop her own ways of utilizing the capacity of post-structuralist theories for exploring nuances and for deconstructing structures of meanings (Søndergaard, 2000, p. 61). Thus, the theories presented here are the result of my own efforts to connect theories and empirical material, and to find research techniques helpful in the analysis of real life examples of gender-technology relations as it will be explored in the following chapters.

Here technology is understood in line with the social construction perspective, emphasizing the co-construction of gender and technology, and also the importance of seeing technology as including symbols, norms, skills and routines, as already described. For theories about gender, my background in history led me to Joan W. Scott's extensive writing on gender and history. Through a number of articles and books, Scott has demonstrated that gender is a social construction, and that its formation as a discursive structure gives meaning to biological sex, or bodies recognized as male and female (Scott, 1988, 1996, 2005). Scott has also illustrated not only how gender, despite being associated with people, also has the ability to 'spill over' and reflect on other matters, activities and objects, and social and political positions, but also how gender has the capacity to appear as 'nature' – as irrefutable and impossible to change (Scott, 1988, p. 60). Other important sources of inspiration for the gender analysis have been Dorte Marie Søndergaard's way of describing gender as a 'sign on the body'. She aims at deconstructing sex and gender into a matrix of different components where the sign on the body is one, and others are related to sexual and professional presentations of the self, illustrating how the individual needs to score high on at least some of the components in order to remain socially recognizable (Søndergaard, 1996). Also, Toril Moi's reading of Simone de Beauvoir has been important to my work, in particular de Beauvoir's emphasis on how a woman defines herself 'through the way in which she makes something of what the world makes of her' (Moi, 1999, p. 72, referring to de Beauvoir's *Le Deuxième Sexe*, 1949). This perspective is particular important in Chapters 3 and 4 in the analysis of how male and female computer students negotiate what it means to be a male or female computer student, balancing between hegemonic narratives and individual expressions.

Michel Foucault's genealogical approach to history, as an alternative to searching for essence or origin, proposes to treat history as a social product, and the task of the historian is to search for how meaning is constructed, maintained and suppressed (Simonsen, 1996; Foucault, 1999 [1971]). Foucault rejects the question of whether something is true or not, and finds a more important question in how matters are constituted as true, or how 'effects of truth' are discursively created. Discourse is a concept used with many different meanings, also in Foucault's writing (Mills, 1997, p. 7). My choice has been to use the concept of discourse as defined in discourse theory, developed by the political theorists Ernesto Laclau and Chantal Mouffe in their analysis of political debates and of how meaning is constructed in society (Laclau and Mouffe, 1985). The main concepts borrowed from discourse theory include discourse, chain of equivalence, moment, element and subject position. Laclau and Mouffe refer to discourses as temporary truths on limited areas, for instance the discourse on computing. They do not claim that nothing exists outside discourse, but it is only through discourse that we can access and provide meaning to the material world. In a Derridean sense, the discourse has no centre of its own – there is no single meaning from which the discourse springs (Laclau and Mouffe, 1985, p. 112). Jacques Derrida describes a discourse as 'a differential system in which the absence of a transcendental signified, in terms of a privileged centre, extends the play of signification infinitely' (Derrida, 1978, p. 280). Instead of a fixed centre, a discourse is built upon moments, the meaningful discursive components, tied together in chains of equivalence. The chain of equivalence refers to how all the included moments have been adjusted to the discourse, and through this adjustment, inappropriate meaning, meaning that disturbs the dominant meaning in the discourse, has been excluded. A nodal point, also described as an empty signifier, is the most significant point of a discourse. It is empty, not in the sense of 'without content', but rather without a clearly defined content of its own. Instead, the nodal point is continually constructed through the chain of equivalence. The effect of this homogenization is that the discourse achieves an inner coherence that makes it appear as natural and 'given', and thus a stable and objective truth. The meaningful parts *not* included in the discourse Laclau and Mouffe refer to as elements, belonging to the field of discursivity, reminding us that they are still subject to the discursive logic (1985, p. 111). Articulations are those social practices that contribute to placing elements in new relations to each other, so that their meaning is being altered (Laclau and Mouffe, 1985, p. 105). Thus, simultaneously with a discourse adopting the semblance of being coherent and stable, the field of discursivity and the not-yet-included elements will always threaten to overturn or change the discourse. Finally, the concept of subject position refers to discursive positions that define possible frames of action for the individual, and therefore focus on the interaction between the individual and the discourse. Subject positions thus point to the process de Beauvoir emphasizes in her claim that we become women (or men) through 'what we do with what the world does to us'. While subject positions are purely discursive constructs, the analysis will illustrate how real people deal with discourses and subject positions in ways that have very real consequences for their lives.

Following Foucault's proposal to search for effects of truth rather than the essence of history, this book will explore how effects of truth are created in the discourse of ICTs, and we will see how individuals constitute their own truths about themselves as gendered individuals within a gendered discourse of computing. We will see how male and female computer students experience the effects of the discourse as real, both for themselves and for their environments, and we will see how the discursive logics makes certain stories count, while others remain discursively invisible.

Studying change in gender-technology relations

Development of ICTs during the last century has made many researchers write about the social changes that accompany this development. One of the most impressive accounts of changes in the information society has been delivered by Manuel Castells. His trilogy on *The Information Age* (1996), and also *The Internet Galaxy* (2001) and other texts where he analyses the 'network society', have made him one of the most quoted cyber culture theorists. Other important theorists that are often quoted in writings about modern society are Anthony Giddens with his notion of the postmodern individual's reflexivity (1991), Zygmunt Bauman with his related notion of liquidity of modern society where predefined categories no longer are the one and only guideline for our lives (2000), or Ulrich Beck with his descriptions of today's society as a risk society (1992). However, none of these otherwise fine theorists deals properly with gender, but rather exemplify how research not specifically sensitive to gender appears as general accounts of society. The Internet Galaxy described by Castells, we might all agree, is not the same for the Northern and the Southern hemispheres, but it might also differ for men and women of the Western world, as access to economical, social and cultural resources are not equally distributed between men and women – not even in Norway.[13] And as Christine Skelton points out, Beck's notion of the 'individualized individual' as 'removed from the constraints of gender' (Beck, 1992, quoted in Skelton, 2005, p. 324) in a reflexive modernity is based on a 'limited notion of gender power dynamics' (Skelton, 2005, pp. 323–4). Thus, there are a number of interesting mainstream studies of change in the information society, but none that also deals thoroughly with gender through gender-sensitive theories.

Since the 1980s we have seen a growing body of research on gender and information technology, most of which follows Haraway's insistence on situated knowledge rather than aiming for the 'big stories'. As we have seen, a number of these studies identify changes in gender-technology relations, but do so mostly through 'still images' from a specific context, time or event. A number of studies of women's contributions in computer history have revealed extraordinary women's participation in what is normally told as a story about men.[14] Thus, remarkable women – such as Ada Lovelace, who made the first sketches for a computer program, and Admiral Grace Murray Hopper, who made a number of early and important contributions to computer programming and programming languages – have been lifted out of oblivion (Gürer, 2002). We have seen pioneers such as the ENIAC women who in truly pioneering work programmed the precursor to the proper stored-program computer by hardwiring it, based only on diagrams of its logic circuits, being acknowledged for their work many years later (Misa, 2010a) when computer programming had turned into a risk sport for male computer nerds (Turkle, 1984). Historical studies of women and computer technology therefore focus most often on the most extraordinary women, those who played some major role in computer history, and less often on the gender-technology relation itself as it concerns people in general. Historical records of gender and technology rarely discuss the most recent period in which computer technology has come to make a difference in most people's lives in the Western world, which is the focus of this book.

Neither is the aim here to write the 'big story' either, but rather to deal with the subject of change over time focused on the period after the introduction of the personal computer. The first case study, presented in Chapter 2, focuses on change by exploring how the personal computer entered Norwegian culture in the late 1970s/early 1980s and how it was culturally appropriated during the ensuing three decades. The cultural appropriation will be studied through public discourses in a newspaper and a computer magazine, where ideas about computers were presented, discussed and negotiated. We will see two clear shifts in perceptions of gender and computer technology, from an ambivalent period in the early 1980s to a consolidation of a masculine discourse during the 1980s and 1990s, and further, towards a new recognition of varied groups of ICTs users after the millennium (Shirky, 2008).

While Chapter 2 focuses on the discursive development in society's perceptions of computer technology in general, Chapter 3 focuses on the field of computer education – the field that has been, and still is, perceived as most problematic in relation to gender equality and ICTs; it is also the field that seems to have caused most instances of the 'stability argument'. This chapter has a historical perspective, focusing on how perceptions of gender-technology relations have developed in educational institutions in this field since the early 1980s. Only very briefly in the early 1990s could we find optimistic reports about how the number of women in technological education – in particular engineering – had increased markedly, both in Norway and in other Western countries (Sørensen, 1991; Hayes, 2010a, p. 29). However, this was short-lived, and subsequently we have, with a few exceptions, mainly seen repetitious claims that despite efforts to recruit women, the field of computing is still marked by a gender gap. In Chapter 3 we will explore how this field of education itself has changed since its origin as a subject in its own right at Norwegian universities in the late 1970s, and how the low proportion of women and the question of gender equality have been dealt with in educational institutions. There are some notable changes in how the low proportion of women has been treated in computer education, based on different ideas about the relationship between gender and computers, and we will see a discourse disregarding gender, a masculine connoted discourse supporting the hegemonic image of boys and men as the real computer literates, and a feminized discourse inviting women with regard to specifically feminine qualities. These various perceptions of gender-technology relations have made differing strategies for dealing with the low number of women in computing seem equally plausible. However, despite their differences most strategies seem to have preserved gender as a differentiating category, rather than challenging it.

Over the same period during which we claim the number of women who study computing has decreased or stagnated, we have seen a growing number of alternative computer educations develop within faculties of social sciences and humanities, which is the focus of Chapters 4 and 5. Computer education within social sciences and humanities attracts a larger number of women than the traditional 'informatics' and 'computer science' (cf. Mörtberg and Elovaara, 2010). This development is, however, hard to see if we only focus our research lenses on one specific male-dominated branch of computer education. Chapters 4 and 5 are based on a case study involving men and women participating in a university-level programming course at a faculty of humanities. The students were followed during their first semester at this course, and their perceptions of gender and computers, and their ways of dealing with these perceptions, were explored through interviews, observations and email communication. We will see differences between men and women, and also differences between women and between men in how they construct their gender identity in relation to computing. This case study was carried out at the end of the second period identified in Chapter 2, the period when the masculine discourse grew stronger. We will, however, see variations that indicate that this is a period of change where some of the findings point backwards to the digital gender gap that worried researchers and politicians alike throughout the 1990s, while others point forward, into a new millennium where women to a greater extent have acquired – and been acknowledged to have – close bonds with computer technology. Thus, in Chapter 4 we will meet women who insist on not developing skills because they are women, women who explore computing as a 'forbidden masculine field', and women who reject that there are any differences between men's and women's abilities and opportunities to work with computers. In Chapter 4 we meet all the students, both male and female, and Chapter 5 focuses on the women's individual stories of change; on how they changed their ideas and perceptions, not only about computing, but about their own relationship to computers during the programming course. All the women had previous experience with computers, but even so, most of them defined themselves as 'novices' at the start of the term. They developed their knowledge, skills and even feelings of confidence with the computer during the term. What puzzled me most with these women was the enormous joy and pleasure they expressed in this process as they acquired 'expert knowledge' in a field they previously had not enjoyed, or had not expected to master as well as they eventually did. These women's individual stories of transformation give some illustrative examples of how computers can best be understood as not simply physical technical artefacts, but as part of the sociotechnical web that also includes knowledge, skills, cultural and symbolic images, and daily routines.

Chapters 2 to 5 thus discuss change and variation in gender-technology relations – or, more specifically, gender-ICTs relations – in various ways and on different levels. In Chapter 6 the different stories of change and variation will be drawn together in a diffractive reading (Barad, 2007), focusing on doing (West and Zimmerman, 1987) and undoing (Butler, 2004) gender and technicity (Dovey and Kennedy, 2006). As the research presented in the various chapters was carried out over a period of 10 years, in different settings and with different focuses, it is necessary to draw up a fresh theoretical frame for this final discussion. The 'diffractive methodology' (Barad, 2007) invites a reading of different texts through each other, and this is the methodology used to explore patterns of similarities and differences across the studies presented in Chapters 2 to 5. The perspective of 'doing gender' presented by Candace West and Don H. Zimmerman in the late 1980s pointed to how we are 'doing gender'. In an effort to understand the many new, non-hegemonic, or even deviating forms of femininity and masculinity, more recent gender theories have suggested other ways of seeing gender constructions, for instance, Judith Butler's *Undoing Gender* (2004), which attempts to explain gender identity that is not in harmony with the hetereronormativity of modern society. I would not go so far as to claim that women who have an intimate relationship with computers are deviant; however, in certain ways they are not conforming to traditional gender stereotypes, in that 'they upset a widely accepted sense of order and meaning' (Powell et al., 2006, p. 690, referring to Cockburn, 1985). Thus, theories aiming at understanding gender performance out of tune with heteronormative stereotypes can perhaps help to shed light on the gender constructions going on when women become computer experts or fall in love with the computer, or when men hide themselves to avoid being exposed as not being computer literate. The concept of technicity points to how technology is involved in the formation of our identity, and this concept helps to explore men's and women's access to and recognition of a close relationship with ICTs. Chapter 7, which concludes, considers how the exclusion of women has been perceived, asking how recognized exclusion mechanisms and popular inclusion strategies can be understood in light of what we know about gender-technology relations and its development today, before closing the book with some reflections on the role of researchers in feminist technoscience.

2
Changing Images of Computers and Its Users since 1980

Introduction

Before World War II, the term 'computer' was not the name of a machine, but referred to a human calculating mathematical tables, and many of these human computers were women (Campbell-Kelly and Aspray, 2004; Grier 2005). The first electronic computing machine, ENIAC, built for ballistic calculations during World War II, was programmed by six women who were pioneers in the field of programming (Gürer, 2002; Misa, 2010a). Computer programming was not considered a real profession at that time. It had not yet morphed into the risk sport of young computer-fascinated men, which we know from later descriptions (Weizenbaum, 1976; Turkle, 1984). Instead, women 'were often stereotyped as being good candidates for programming', Denise Gürer writes, linked to abilities such as patience, persistence and an eye for detail (Gürer, 1995, p. 47). The ENIAC women enjoyed their work, but did not see themselves as pioneers, according to Gürer (2002, p. 119). They were simply doing a job, as Kay McNulty Mauchly Antonelli recalls: 'We just thought we were doing the job we had been hired to do' (Mauchly Antonelli quoted in Gürer, 2002, p. 119). And women could perform this, Jennifer Light claims, because it was seen as a job 'at the intersection of scientific and clerical labor' (Light, 1999, p. 457), 'scientific' associated with men, 'clerical' with women. Also in the following decades, women were important in the computer industry, as pointed out by Marlaine Lockheed, claiming that when this industry 'was young and there were only 2,000 computer operators, 65% of them were women' (Lockheed, 1985, p. 117). Later, when the computer entered offices in large numbers the US Bureau of Census assumed that a higher proportion of women than men were computer users in the US labour force (Ahuja, 2002, p. 20). Why, we could ask, did programming turn into one of the most hard-core male fields of computing, and why did computers and computing become 'a masculine world' (Mahoney, 2001, p. 171)?

Traditionally this story has been told as a story of men and their machines (Misa, 2010a), with a focus on male engineers, hobbyists and hackers, as well as male-dominated institutions. As in many other fields, women's contributions remained more or less hidden[1] until historians started focusing on women in particular (Abbate, 2003), and documented how, also, women '[f]rom the very birth of computing machines... made substantial contributions' (Gürer, 2002, p. 116).

Pioneering women in computer history, such as Ada Lovelace,[2] Grace Murray Hopper[3] and the ENIAC women, were in certain ways extraordinary, or put in extraordinary positions in relation to computers. In fact, most people would not see, and far less touch, a computer for yet another three or four decades (Ceruzzi, 1998; Campbell-Kelly and Aspray, 2004). Looking for changes in a long-term perspective makes it obvious that both men's and women's relationships to computers have changed, as both computers and society in general, including gendered norms and regulations, have changed. However, we are not taking on a long-term perspective in this chapter, but will instead focus on the more recent period of computer history: the history of the personal computer since the early 1980s. The small desktop computer, quite different from earlier computing machines in their various types, has come to affect more people more directly than the large computers of the 1950s and 1960s. Changes might seem more diffuse and perhaps harder to unravel for this later period, as the connection between boys and men and computers has come to seem obvious, as 'an academic "truth"' (Grint and Gill, 1995, p. 3) as well as a historical fact, in more recent decades. Jörgen Nissen's article from 1996 titled 'Of course it is boys who do stuff with computers! But why is it so?' [Det är klart att det är grabbar som håller på med datorer! Men varför är det så?] illustrates that the connection between boys, men and computers along the way – between the female human computer and the modern computer – has been taken as obvious. Nissen also illustrates our search for explanations as to why the computer turned out to be something for boys and men. It is far easier to document the close connection between men and computers than to find solid answers to this development. The hegemonic character of the masculine discourse of computers makes it 'obvious' in such a way that it becomes invisible as a social construction. There is, however, no 'natural' or essential link between masculinity and computing, and thus it is important to ask how computer technology became a masculine field.

Researchers have tried to answer this question, for instance, by pointing to the development of computer technology within male-dominated fields, such as the military and universities (Mörtberg, 1994). But does this really explain why computers have continued to be perceived as a masculine technology? What if we change the perspective? Instead of focusing on those who developed the technology, we could focus on its qualities, or the way it has been used. When the personal computer first started to invade the market in the late 1970s and 1980s it 'assumed a shape... that tied it to cultural icons and activities that in and of themselves seem less straightforwardly masculine', Michael Mahoney points out (2001, p. 171). Instead, the computer fitted an image of women's work tasks. It was about handling a keyboard, inheriting the typewriter's place in office work, and it matched feminine qualities, such as nimble fingers. 'Computing could have been gender-neutral', Wajcman claims (1991, p. 150). Gürer tells the story of one of the women computer pioneers, Judy Clapp, who had managed to convince her male boss to get a computer: 'He got his computer and called me to come up and see him using it. I walked in and he was leaning back on his desk with his secretary in front of the computer. He was dictating and she was typing!' (Gürer, 2002, p. 119). Men did not want to use the computer because they perceived it as a tool for female 'typists', Clapp recalls (ibid.). Perhaps we should be more surprised that the technology did not become a gender-neutral or even feminine technology?

In this chapter we will explore how the personal computer entered and was appropriated in Norwegian culture since 1980. The computer did not represent a new technology around 1980, but even so, something new happened. While, prior to 1980, computers had been a technology for those with a special interest or a special need, the development of personal computers that began in the mid-1970s made computers potentially available to more people and to different groups. Although changes did not happen overnight, the new availability of computers marks a new era for public debates concerning the social and cultural meanings related to the personal computer, in Norway as in other Western countries (Huws, 1991).

Many books have been written about the development of personal computers in the 1970s and 1980s. Some focus on computer history in general, which more often than not means on technical development and market development (Ceruzzi, 1998; Campbell-Kelly and Aspray, 2004). Others focus on how the underground movement developed the personal computer intended for private use, thus developing a market of which the larger producers, such as IBM, had not yet become aware[4] (Levy, 1984; Ceruzzi, 1999). We know less about how these developments affected and were affected by the 'main' culture, or the larger society, or about how the computer was perceived and appropriated as a cultural artefact, or how it spread among 'people in general'. Neither does gender seem to be an important analytical category in mainstream computer history.

The aim here is to explore how the discourse of computers developed as a gendered discourse in Norway since 1980. How have computers been attributed social and cultural meaning that also included ideas about gender? How has the image of men's close relationship to computers, along with ideas about women's lack of interest in or even their distance from computers, developed? Who were considered computer users, and who were not? Thus, this chapter aims to explore how the computer has been *culturally appropriated* in Norway since the 1980s. We will explore how the relationship between gender and computers was perceived and presented in public discourses in Norway. The empirical material is primarily gathered from the newspaper *Aftenposten*, Norway's largest newspaper, but also from the computer magazine *Datatid*.[5] A keyword in the analysis that follows is 'discourse', and we will investigate how discourses about computing and gender have developed since 1980. Discourses often appear to be fixed or stable, but when scrutinized we will find that they are unstable, socially constructed structures, continually changing over time and from place to place. Here, we will explore how the discourse of computer technology developed, focusing on the public sphere in Norway – that is, on debates that were available to and concerned 'people in general'. Thus, this is not 'computer history' in a traditional sense. It is not the history about what 'really' happened, but rather about how the development was *perceived* in the public sphere.

As we saw in Chapter 1, technology researchers have made recurring claims that technology should not be treated purely as physical artefacts, but rather be seen as a social construction, involving knowledge, routines and practices. Social constructivist theories have emphasized that society and technology shape each other in a mutual process (Berg, 1996; Lohan, 2000; Faulkner, 2001). The theory of domestication focuses on the adoption of new technology in the household (Silverstone et al., 1997 [1992]), and emphasizes that a household is not an empty space, but has its own 'moral economy' including already established routines, values and practices. New technology brings something new into the household, but the technology also has to be translated into the moral economy of the household, and so they affect and adjust to each other. It is also possible to see Norwegian culture as an arena for domestication of new technology. New technology brings new elements into a culture, at the same time as the technology needs to be adjusted to that culture, and to acquire meaning in and through already meaningful cultural stories (Pfaffenberg, 1988). We will explore how these discussions of gender and computer technology helped construct the discourse about computers in a particular way, through a series of 'discursive logics' that resulted in a homogeneous and hegemonic discourse in the period up until 2000, creating different expectations about men's and women's relations to the computer. Despite the seeming stability, there have been notable discursive changes since 2000 that will be discussed in the second part of the chapter.

Discursive logics before 2000

Introducing the personal computer

The personal computer is an invention from the second half of the 1970s.[6] Although the computer as such was not new at that time, the advent of personal computers drastically increased the *potential* availability of computer technology for private use. A number of cheap computers became available on the Norwegian market in the late 1970s and early 1980s, some targeting the business market and others targeting the home and hobby market.[7] The new availability of computers did not mean that things changed overnight. In 1985 only 9 per cent of the Norwegian population aged between 9 and 79 years had access to a home computer (Statistics Norway, 2011), but the computer-using population grew steadily to 50 per cent in 1997 and 71 per cent in 2000. The Internet entered the scene in the 1990s, with access increasing rapidly from 13 per cent in 1997 to 52 per cent in 2000 (Vaage, 2010, p. 80).

The computer entered Norwegian culture as a heterogeneous technology, at the intersections between work, home, school and society at large, promising dramatic changes and causing a 'new technological and social revolution'.[8] Media created already in the early 1980s an image of home computers entering Norwegian homes in high numbers, and 1995 was described as 'the year we connected to the Internet'.[9] The early 1980s were seen as a time of change that would affect everything and everyone, from the nation to individuals, regardless of gender.[10] Labels such as 'computer', 'information' and 'technology' were used as futuristic prefixes to 'society', describing the entire society in light of the new technology, where computers made *the* difference (Lie, 1996): 'Our society is about to change – and the "driving force" is the computer.'[11] The computer constituted 'the language of the future', it was claimed, and 'those who do not learn the language of the computer today will be the illiterates of tomorrow'.[12] In numerous media reports, women were explicitly included in these images, assumed to be using the computer in the home for typical household tasks.[13] In the computer magazine *Datatid*, computers were seen as giving women with caring responsibilities in the home a new opportunity to enter working life, because computers made work more mobile.[14] However, this theme disappeared during the 1980s, and women were soon labelled as being at risk of becoming 'the illiterates of tomorrow'.

The revolutionary potential of computer technology was one of the recurring themes in both *Datatid* and *Aftenposten*, and in descriptions of the home this took on a special form, visualizing the computer doing 'everything'.

> Computer technology will offer a myriad of possibilities for people at home and in their leisure time. You can have your own cookbook, medical book, a number of games. When is the tram leaving? Just ask the computer. And what products do you have in the fridge? Most people will probably use the family's computer to keep track of their economy. So far the computer technology has been too expensive and too difficult to use for it to become widespread in the home, but both of these barriers are about to disappear now.[15]

This entry from 1981 describes the future use of computers, but a future that was not far away. The computer was just about to 'enter the living room', albeit the living room of people not familiar with the computer. Consequently, a market and a need had to be created (cf. Nye, 2004). Huws reports on similar stories in British newspapers in the late 1970s and early 1980s with 'a curious mixture of doom-laden prophecy and boyish excitement', including stories about the 'home of the future', about a 'marvellous range of activities which could be carried out "from your own living room"', and with expectations about women's new opportunities in working life (1991, p. 23).

Bryan Pfaffenberg claimed that for new technology to become culturally appropriated it needed to acquire meaning in and through already meaningful cultural stories (1988). The same went for the practical domestication of new technology, which, in order to be successfully included in everyday life, somehow needed to support what appeared to be meaningful activities. Listing the contents of the fridge was not one of them. Thus, the attempt to enforce a domestication of the computer in the home was relying on creating a 'need'. In this process, the computer was described as fulfilling a number of roles that it did not fulfil, and fulfilling needs that did not exist, illustrated by a question posed in *Aftenposten* in 1985: 'The computer is the answer, but what is the question?'[16] This construction of the computer as important to a wide set of activities also increased the importance of being computer literate, in order to be able to find the 'recipes for roast lamb or meatballs', to manage your personal economy and to resolve other matters of greater or lesser importance on the computer. Thus, this also enhanced the signification of the gap between those that were computer literate and those (mainly female) that were not, and the cultural importance of the new computer technology seems to have anticipated its actual diffusion.

These examples illustrate that computer technology was about to acquire a place in culture, a place that was not pre-defined, but that had to be created in and through a collective process of domestication. During the 1980s, the computer was frequently discussed in relation to work, school and education, and to all age groups. The discussions about the relation between society and technology concerned contexts and groups that involved both genders. In most cases, gender was not explicitly mentioned. Here, however, we are focusing on the cases where gender *was* perceived as an interesting category, and we will now see how certain discursive logics worked to establish an image of computers as being closely associated with men.

A pattern of visibility and invisibility

The most common focus on girls and women in *Aftenposten* concerned their lack of interest, experience and skills, while by contrast the dominant focus on boys and men was their fascination and extraordinary computer skills. This trend contributes to making computer skills both visible and invisible in a certain gendered discursive pattern.

Aftenposten writers focused on women's lack of interest in computers and their low level of participation in computer-related activities, at work, in education, in the private sphere and in society in general. Statistics showed a gender gap in access and use,[17] and choice of education,[18] as well as referencing situations where girls and women were less engaged than boys and men (in the classroom,[19] at computer parties,[20] at home,[21] in school,[22] at work,[23] on the Internet[24]) or in general a problem for society.[25] All pointed to women's low level of participation. Boys and men were often mentioned in these entries, representing the 'norm' against which girls and women were measured and found wanting.[26] These entries increased the visibility and discursive importance of female non-users, and women's low level of interest in computers had become 'common knowledge' by the late 1990s[27] (Corneliussen, 2003a; see Chapter 3 in this book). Not merely the large number of such news reports, but also the descriptions, comments and evaluations in many of the reports focusing on women's lack of interest increased the discursive importance of women as non-users. A typical entry was titled, bluntly, 'Women do not give a toss about computer technology.'[28] Norway was presented as unique since 'nearly all Norwegian homes have a PC' (surely an exaggeration of the real situation) and since a larger proportion of the Norwegian population than in other countries had access to the Internet. However, two surveys documenting women's low level of use were quoted. First, a survey from 1997 stated that only 1 per cent of women who had access to the Internet actually used it, against 12 per cent of men. And then a survey from 1996, showed that only one in five women as opposed to one in three men used the PC at home.[29] The surveys obviously documented a lower level of usage among women. This focus on the difference *between* men and women strengthened the impression of women as non-users, not only by ignoring the female *users*, who – although not representing a large proportion at that time – still existed, but also by disregarding the fact that a *dominant majority* of men were also non-users, with two thirds of men not using the PC and 88 per cent not using the Internet at home. The male non-users were simply not discussed at all, not in this or any other news report presenting statistics. Women were identified as the future losers, while a similar threat towards male non-users was overlooked. But why would it be more interesting that women were lagging behind compared to male Internet users, rather than the fact that the majority of men were also non-users? And why would only *female* non-users face the threat of becoming 'the losers of the future'?

This episode illustrates how relations between gender and technology were made visible and invisible in a particular pattern. Women's non-use and men's embracing of technology, even though it did not include a majority of men, were made visible. Simultaneously, male non-users and female users were ignored or made discursively invisible, contributing to an overall homogenization of a masculine discourse of computers.

Not only were women labelled as indifferent to computers, but they were also associated with technophobia. A researcher who described technophobia as something that had always existed was asked by a journalist: 'Isn't this [fear of technology] something that in particular applies to women?'[30] The researcher answered the question without responding to the gender element. Although the question about women does not appear to have meaningful significance in the interview, the journalist decided to retain the question – more precisely, the assumption about women – in the published text. Even though the question was not reflected in the dialogue that followed, it did *create* meaning in the text by connecting women and fear of technology, illustrating and reinforcing the readily available connection between women and technophobia. And, it was allowed to stand unchallenged (imagine the journalist asking the same question about men and the interviewee not responding!). Thus, the reports focusing on women's lack of interest in the computer greatly enhanced the impression of women as non-users, making them discursively visible, at the cost of female users, who remained discursively invisible. Similar invisibility of women in relation to computers was also found in other Western countries, as illustrated by Lockheed. Writing in 1985 from the US, Lockheed is fighting the image of the 'computer hacker' who 'clearly permeates our conception of computer users as insensitive rule-oriented males with a lust for winning', while emphasizing that women also are computer users (p. 116).

In sharp contrast to the treatment of girls and women, the dominant focus on boys and men in the Norwegian newspaper (*Aftenposten*) and computer magazine (*Datatid*) was rather their exceptional interest, extraordinary skills[31] and even their love for the computer.[32] Two recurring themes operated in parallel. First, the reports obviously admired the boys' and young men's skills. They were described as self-educated computer 'wizards'.[33] Young men in computer companies were 'dragged out of the boy's room',[34] to work in a business 'where the geniuses are so young they barely escape the penal code against child labour'.[35] They belonged to the 'hacker generation'[36] and their only education was what they had taught themselves in 'the boy's room'.[37] They 'lived for their work'[38] and had 'the fiddling in their fingers and the computer technology in their heads', talking in a technical language far beyond anything even the friendliest mother could possibly understand.[39] These young boys were the masters of the language of the future,[40] as well as the future leaders of society.[41]

The second recurring theme concerned the negative aspects of these male-dominated computer activities: illegal copying of computer programs – in particular games,[42] violence and sex in computer games,[43] and the negative effects of playing the games.[44] With the advent of the Internet, online activities generated further worries in the shape of computer viruses, 'starting as an innocent boy's game';[45] online hacking of computer systems;[46] and violence, sex and pornography on the Internet.[47] Criticism of the boys' activities did not, however, change their image as highly-skilled computer wizards. In most of these reports 'the boy's room' was an important metaphor for the computer-skilled boys and young men – even in the computer industry, where offices resembled a boy's room more than a workplace.[48] The metaphor became so important as a norm for computer skills that one of the universities even offered a special 'boy's room competence course' for female computer students, 'in the things we assumed that the boys knew about computers before they attended the program'[49] (see Chapter 3 in this book).

There are striking differences between the news reports focusing on women and those focusing on men. First, reports discussing women often involved a comparison of women with men, while entries about men most often remained focused on men alone. Second, women's attitudes towards computers were often explained through analysing them 'as gender', while gender rarely figured as an explanation in articles about men (cf. Faulkner, 2000b). Third, the reports about women frequently encouraged women to acquire computer skills, while not a single report in *Aftenposten* encouraged men who had not yet become computer users to follow suit. Finally, while the focus on women's lack of interest made female computer users invisible, the focus on computer-skilled men made unskilled boys and men invisible. The atypical self-made computer wizards from 'the boy's room' were deemed to represent the 'typical' male relationship with the computer (cf. Sørensen, 2002b).

This discursive construction also seems to have made it easier to 'remember', and consequently to repeat, gender as a feature assumed to make a difference. Several news articles presented statistics showing a difference between users and non-users related to gender, age, education, income or geography. Sometimes all of these features were involved, sometimes only a few of them, and occasionally gender was explicitly mentioned *not* to be one of them.[50] However, gender, and to some degree age, are the only markers that were commonly discussed as important. Thus, the hegemonic discourse made certain meanings suitable and others unsuitable, and it was apparently easier to rely on gender as the primary difference rather than other social characteristics such as class, education or age. An entry discussing use and non-use of different media in Norway quoted a survey showing that 'technophobia' was mostly found among women, the elderly and the poorly educated. The (male) author added a postscript: 'Age is not an important factor'.[51] The rejection of age as important illustrates a homogenization of the discourse by neglecting the 'unsuitable' differences, making the remaining differences even more important.

The almost exclusive focus on girls and women as lacking interest compared with the focus on boys and men as being highly-skilled computer users produced a particular discursive effect. Male users and female non-users were made visible and included as 'self-evident' in the hegemonic discourse, whereas male non-users and female users remained invisible and were in effect excluded from the discourse. This pattern of inclusion and exclusion – based on making computer skills and gender visible and invisible – is perhaps the most important construction in this discourse, and it also forms an important basis for the next discursive construction: the 'intersection rhetoric'.

The 'intersection rhetoric'

The dominant perspective, assuming the computer would have revolutionary impact on all societal spheres, fostered a distinctive 'intersection rhetoric'. It featured a tendency to use one context, such as home or work, to 'explain' a completely different context. One example is found in a report discussing 'computer phobia' in working life.[52] Although computer phobia was found among both men and women, it was more widespread among women, or so it was claimed, illustrated by a woman describing her difficulties in using a computer at work. This report went on to explain the differences between men's and women's relations to the computer by referring to a study of computers in Norwegian households, where many women seemingly resisted the computer because of their husbands' intense and time-consuming use of it. Thus, women's resistance was explained as a protest against men's extensive use. However, in the context of this news report, research about women's resistance to computers in the *home* was used to shed light on women's computer phobia in *working life*. The computer's position at an intersection between different fields of society developed into a particular intersection rhetoric, making observations or arguments from one sphere appear to be valid for another sphere.

Another example of this intersection rhetoric starts with an 'expert' from a computer company who worried about women's inability to keep up with the technological development. 'Many women see the PC as a boy thing for men who are a bit childish', he claimed, and too many women are spectators to, instead of participants in, the rapid development.[53] In consequence, women were facing the risk of becoming 'the losers of the future' who would miss out on exciting jobs and might even 'have to go back to the kitchen sink or low paid professions'. They would also 'experience a greater distance from children's everyday life because they do not understand what children are doing'.[54] The intersection rhetoric in this report involves education, working life and the home, as well as different levels of computer skills. One university lecturer claimed that it was often 'required that you can handle a PC when you apply for a job', then moved on to talk about the small number of women in computer science, again illustrating the intersection rhetoric by conflating two fundamentally different types or levels of computer knowledge.

The fact that the computer entered culture at an intersection between different spheres and different levels of users, from the hobby user to the secretary to the expert, apparently made it easy, and even natural, to use an intersection rhetoric, where arguments from one sphere or one group of users could be applied to another sphere or user group.

Non-hegemonic men and women

The dominant or hegemonic discourse presented men as computer-skilled and women as disinterested. However, there were two additional groups in this period, a group of female users and a group of male non-users. How do these fit into the hegemonic discourse?

In several reports, women were actually portrayed as the 'super users' or the main user group for office computers, both in the private and public sectors.[55] 'Secretaries have in many places become the companies' leading lights in computing.'[56] It was noted that 'many female secretaries do the computing for their male leaders',[57] and many of them 'are more competent than their superiors with regard to computers and technology'.[58] Thus, it appeared, the secretary was not 'automated' out of a job.[59] Instead, the secretary's job was re-evaluated and transformed into a new, though still female, 'office manager' with a more responsible job, higher wages and higher status.[60] Together with observations about women being well-represented or dominant in low-level computer work, such as routine work and word processing,[61] these reports acknowledge women as an important user group for office computers. Women were also important users of office computers and in routine jobs in other Western countries (Lockheed, 1985; Ahuja, 2002). According to Thomas Haigh women made up 95 per cent of keypunch operators in the US in the early 1980s (2010, p. 64). Nonetheless, despite the clear acknowledgement of their computer skills, these female office workers did not alter the hegemonic discourse on computer use, which remained focused on female non-users. This important group of female computer users could in effect be ignored owing to the perceived difference in the 1980s between a secretary's computer, deemed a mere 'word processing machine', and a 'general' computer (Lie et al., 1984, p. 25), together with how women's use was limited to operating the technology. Using male-dominated computer expertise as the norm for 'computer competence' would wipe this (female) group off the map. These women computer users were made visible as women, but in the hegemonic discourse their computer use did not count for much.

The hegemonic discourse also effectively ignored the male non-users, or 'the Stone Age leaders', as they were called. In the computer magazine *Datatid*, this curious group was quite prominent throughout the 1980s and 1990s, but they were invisible in *Aftenposten* until the mid-1990s. In 1995 *Aftenposten* profiled a number of 'renowned' non-users – all of them in high positions, and all of them men. One of them, apparently not so happy with being interviewed, stated: 'I think of myself as smart enough to master this, but with so many computer competent people around me I do not want to spend time on it.'[62] This claim echoes an informal survey among male leaders in Norwegian computer companies undertaken by *Datatid* in 1985, which showed that the leaders appeared busy with other things and so left the computer to be handled by the secretary (Corneliussen, 2006). These male leaders presented their non-use of computers as a *deliberate choice*. This choice was apparently accepted, at least until the end of the 1990s. A journalist writing in *Aftenposten* in 1999 acknowledged that he recently started using email after being rebuked by a business associate: 'Do you live in the Stone Age!'[63] This article made fun of leaders and people in high positions who did not use computers, especially those not using email. Another non-user would tell people that he had a PC in his office, followed by a whisper: 'But I don't use it!' Many of these male leaders got a PC in their office in the early 1990s, it was claimed, because 'it looked pretty', but many of them still wrote by hand.[64] Apparently some of the male leaders acquired the *status* that followed from being recognized as a computer user quite early without actually using it. The consequence of their non-use was described as no more severe than the risk of receiving business communications on paper through 'snail-mail', as before, a far cry from the dire consequences described for women non-users.

These prominent men were obviously not the only male non-users in the 1980s and 1990s, if we recall that only 9 per cent of the Norwegian population had access to a computer at home in 1985, growing to 50 per cent in 1997 (Vaage, 2010, p. 80). Still, they were the only group of men made visible as non-users, both in *Aftenposten* and *Datatid*. Despite the leaders' visibility as men, they were not discussed 'as gender', and their gender was not used as an argument or explanation for their behaviour. They did not affect the general image of men as computer competent, but were instead excused on the basis of a different discourse – the discourse of the busy business leader.

Challenging the hegemonic discourse in reconstructions and new voices

So far we have seen how the hegemonic discourse was created through a systematic pattern of making female non-users and male users visible, while simultaneously male non-users and female users remained discursively invisible. Intersection rhetoric increased the visibility of women's lack of knowledge by permitting the use of observations or arguments from one arena, group or level of use to be valid explanations in another arena, group or level of use. By this overemphasis of gender, a conception of gender as the primary difference was sustained, and differences within the genders as well as other differences, such as age or income, were made seemingly unimportant (cf. Sørensen and Nordli, 2005). Even when male non-users entered the scene, they did not face the threat of being excluded from important activities. And when the hegemonic discourse was occasionally challenged, it was met with doubt, disregard or even contradiction. These discursive logics all worked to homogenize the discourse by including meanings that supported the hegemonic discourse, while excluding meanings that threatened it. However, it is important to note that discourses do change, despite their seemingly stable and fixed qualities. In this section we will see attempts to rewrite the discourse before 2000, as well as the entry of new voices and new meanings after 2000.

How to change things if gender is a primary difference

One of the recurring questions following the worries about women's low interest in computers, particularly in computer education, has been the question of *how* to include women (Björkman et al., 1997; Faulkner and Lie, 2007). The trend in computer education in Norway has to a large degree followed the rest of the Western world: the low proportion of women grew slowly during the early 1980s, but decreased markedly towards the mid-1990s (Charles and Bradley, 2006; Hayes, 2010a).[65] Numerous initiatives to recruit and retain women had temporary and local effects in the late 1990s (Lagesen, 2003; see also Chapter 3 in this book), but the numbers of women once again decreased after 2000, and continue to decrease for the most technical computer subjects.[66]

In Chapter 3 we will see how the gender question in computer education and recruitment initiatives has been treated along different lines within the educational institutions, from a 'gender-blind' to a 'masculine' and a 'feminized' discourse of computing. In this chapter we will concentrate on how the 'girls and computing problem' (Gansmo, 2003) was treated in the press; there are traces of the same discourses in *Aftenposten*. We find the gender-blind discourse blaming women for making the wrong choices,[67] and the masculine discourse encouraging women to 'think in new ways'[68] or to 'close the dairy and the pedicure parlour and start up in the computer business'.[69] We even find a more feminized version trying to rewrite the meaning of computing as less technical and more social – in accordance with women's supposed 'social abilities', similar to findings by Alison Phipps in her study of discourses about women in science, engineering and technology (SET) in Europe and North America (2007). While the first two discourses tried to change women, the third is the most interesting in this context, as it tried to change the discourse of computing to make room for women, and this is the one we will focus on here.

Reports in *Aftenposten* discussing women's low participation in computing education and low interest in computers frequently tried to explain this fact, as well as how to change it, by conceptualizing men and women as essentially different. Men and women have different attitudes towards computer technology, it was claimed. It was assumed that men have a playful and exploring attitude while women are driven solely by need, not by enthusiasm or inquisitiveness.[70] Men get addicted, and they love the technology for itself, whereas women ask what they can use it for. Boys are interested in 'finesses and technique' while girls are interested in 'communication, email and information retrieval',[71] arguments well-known from earlier research both in Norway (Aune, 1996; Håpnes and Rasmussen, 2003) and in other Western countries (Turkle, 1984; AAUW, 2000; Gannon, 2007). Men and women were also described as having essentially different qualities, and several entries claimed that women in certain ways were *better* than men with regard to computing: 'women are better in comprehending the user's situation. Men have a tendency to lose themselves in exciting details.'[72]

A contemporary campaign echoed these very arguments to recruit women to computing. Here, women were lauded for their excellent communication abilities, while men were acknowledged as 'the technical geniuses' (cf. Chapter 3 this book; Corneliussen, 2003b; Lagesen,

2003). This description of men and women coincided with an attempt at redefining computing. One business representative claimed that many computer jobs were 'not technical at all', and another expert claimed that 'the characteristics associated with girls...creativity and to be good at teamwork'[73] would lead to success in the computer business. The arguments were perhaps well-intended in the way they made room for women, and described computing as something women could manage as well as or, in some instances, better than men. However, the underlying argument was that computing was not *really* about technology, but about something else – reading between the lines, that women were not *really* interested in the technology, but in something else.

Emphasis on the social aspects also surfaced in a rare description of a group of female computer engineers. They were described as experts, but as a negation of male computer experts: they did not talk incessantly about technology and they were certainly not nerds. 'Instead they talk about weddings, children, men...about a lot of things that often occupy women in their late 20s and early 30s.'[74] The female computer engineers emphasized that they 'worked with people' – just as they assumed other young women would want. Besides, computer companies did not always want male computer geeks, it was claimed, but rather preferred women who could 'listen to users and their needs'. Thus, the message was that the computer industry needed women because they were *not* like male experts. Still, it seems, they were not valued for their technical expertise.

A similar emphasis on women as people oriented is also found in other Western countries (Phipps, 2007). Faulkner has pointed out that this is part of a dualism 'between "the technical" realm and "the social", by which men/masculinities are so readily associated (symbolically) with technology and women/femininities with people' (2006, p. 6). This dualism is mutually exclusive, thus 'to be technical is to be not social and vice versa', and Faulkner suggests that this might be 'a key reason why engineering is often perceived as a "gender inauthentic" option for women' (Ibid.). The suggestion that computing is incompatible with femininity (cf. Turkle, 1988; Wajcman, 1991) also creates an imbalance that may lead women to maintain a distance from male computer experts in attempts to restore the balance. Alison Adam et al., in their study of women in the UK IT workforce, illustrate that many women find it difficult to find a balance between being a woman or being perceived as feminine and their professional identity. They point out that 'it is notable that these often highly technically skilled women play down their technical knowledge' (2005, p. 288). Ruth Woodfield supports the view that 'women as a group, by dint of their socialisation' are more people-oriented than men as a group (2002, p. 121). However, Woodfield has also pointed out that women's social and communicative skills are explained in terms of 'gender-typed personality characteristics' which make them less important as 'skills', but rather associated with women's nature. This is different from descriptions found by Woodfield, of men who were perceived as having good social and communicative skills along with their technical skills. Woodfield's informants rather described these men as 'exceptional', 'super-clever', and as having an 'equally advantageous, if not superior ... male approach to working with people' (2002, pp. 127–8). Thus, the discourse of women's social and communicative skills is challenged by men showing similar 'if not superior' skills, simultaneously as women portraying these skills risk being perceived as 'too careful, and sometimes too "caring"' (Woodfield, 2002, p. 128). In addition to how the special characteristics attributed to women involve a potential risk of retaliation or their capture by men in superior versions, Schiebinger has also pointed out that the gender characteristics often attributed to women emphasising social skills, cooperation and caring and so on 'date back to the eighteenth century and were produced in efforts to keep women out of science and the public sphere' (Schiebinger, 2008b, p. 15).

As far as we can see, women confront a number of 'double standards' in computing. They face difficulties in establishing an identity as a 'computer expert' in part because they lack ready-made images or gender-authentic role models with which to identify. Re-writing computing to be more in line with the assumed 'female qualities' and preferences offers an alternative co-construction of gender and computers. However, not everyone is entirely happy with this attempt to write technology out of computing; neither are they happy with the negative image it creates of men (Lagesen, 2003). And the alternative construction does not really include women in the technological aspects of computing.

New voices after 2000

The hegemonic discourse seems to have been stable through the late 1990s, while attempts to re-cast the discourse in this period occur primarily with concerns about the low proportion of women in computer education. After 2000, we find new discursive constructions alongside the dominant discourse. Different meanings can exist simultaneously, resulting in something akin to a polyphonic musical choir of different and seemingly contradictory voices (Bakhtin, 1986; Lagesen, 2005; Hayes, 2010b). After 2000, we can find this polyphonic choir with echoes of previous voices and with new voices. The newspaper articles on gender and computing since 2000 can be roughly divided into four main topics: (i) criminal men; (ii) computers for everything and everybody; (iii) differences between computer users; and (iv) women in IT education/business. The articles within these four topics also relate to gender in distinctive ways.

The articles about criminal men are all very brief, as though they were barely worth mentioning. Continuing from the earlier period, it is still men that are associated with criminal use of computers. Yet in this period the young male hacker, formerly the computer wizard of the 1990s, lost some of his status.[75] His criminal deeds are now described as 'boyish pranks',[76] and he is no longer celebrated for his arcane knowledge. Instead, his computer knowledge is 'normalized', since anyone can find recipes for hacking on the Internet.[77] An article featuring a survey made by a security company even claims that among those aged between 15 and 17 two out of three have attempted to hack into other people's online accounts.[78] Thus, the call to 'forget the hacker' (Gansmo et al., 2003a) is partly achieved by redefining him from being someone with exceptional computer skills to an 'anybody' using commonly available information. The young male hackers or nerds have clearly lost their hegemonic position as the celebrated computer wizards, and 'hacking' has become so widespread that it is no longer an activity for boys, but involves a majority of youth, including girls.

One of the most notable differences since 2000 is this embrace of computers as being everywhere, used by everyone and for everything. The earlier period also saw similar claims, in particular during the early 1980s, although mostly referring to the future (Huws, 1991). The more-recent news articles seem to reflect practices already present and realized, and they construct the categories of 'everywhere' and 'everything' more actively, not as a vision but through stories of people's actual use of computers. The category of 'everybody' is actively constructed as a discursive group that includes many *new* users of computers. Thus, among computer users in this period we meet young people, old people, men, women, grandmothers, men and women in their sixties looking for a date, office workers, prisoners and prostitutes; and they use computers for a broad spectrum of activities including work, hobbies, communicating in new ways with family and friends as well as with strangers on Internet services.[79] As we saw earlier, access to computers among the Norwegian population exceeded 50 per cent in 1997, and increased further to 71 per cent in 2000, 79 per cent in 2004, 85 per cent in 2006 and 92 per cent in 2009. Also, Internet access increased rapidly, exceeding

50 per cent in 2000, 66 per cent in 2004, 79 per cent in 2006 and 91 per cent in 2009 (Vaage, 2010, p. 80). In comparison, households with Internet access in the UK amounted to 73 per cent in 2010,[80] while 82.5 per cent of the population in the UK and 77.3 per cent in the US had access to the Internet in 2010, according to *Internet World Stats*.[81]

In 2009, we also find a new category in the official statistics, as it was no longer simply a question of whether people had access to a PC or not, but rather *how many* PCs to which they had access, which was 2.1 in Norway in 2009. ICTs achieved a new level of diffusion in the period after the millennium, and were no longer esoteric, but a natural part of a growing majority of 'everybody' (Elovaara, 2004). The new user group of 'everybody' still confronts images of the 'original' computer experts – that is, the nerdy men. In challenging the 'nerdy men', the most 'unlikely' users are called upon, such as, for instance, 'your aunt', 'your mummy', 'women over 60' – images that clearly challenge the image of the 'real' computer user as solely male (Gansmo et al., 2003b, p. 132). Simultaneously, these new images counteracted the invisibility of female computer users.

Some of the new computer users are presented as marginalized females, such as prostitutes and female prisoners. In their stories, computer technology has gained a vital position in making them able to participate in and become successfully integrated in society. Even though male non-users never really existed in the hegemonic discourse, new male users are also presented in this period, such as male craftsmen. Craftsmen did not consider computers important, but now they realize that running a business without computers is about to become impossible, or at least very embarrassing.[82]

The new spectrum of users challenges two of the most important discursive constructions from the earlier period: the dominant position held primarily by young males as the 'real' computer users; and the image of women as non-users together with the invisibility of female users. Thus, these new stories challenge the gendered pattern of visibility and invisibility or inclusion and exclusion of the earlier hegemonic discourse, creating a new discursive space for both men and women as computer users.

This expansion of computer users clearly involves a number of different users with a number of different social features. The newspaper articles referred to above do not explicitly discuss in which ways they are different. This is, however, the topic in a series of other articles in which gender is again described as the main differentiating feature. The old, pre-2000 voices emphasize that girls have less technical skill levels than boys, and that girls want to maintain distance from the nerd label, emphasizing the gender gap of the hegemonic discourse. In dramatic contrast, the new, post-2000 voices observe that the gender gap is about to close. Women have 'invaded the Internet', with gender differences in access and use 'surprisingly small'.[83] However, women turn out to use the Internet in different ways, since 'contact with friends and acquaintances is more important than tangible results'. Men on the other hand, 'want action and results', banking services and 'window-shopping', while boys are 'most eager to use the net for creating contacts'. Inconsistencies in descriptions of the various preferred activities seem to indicate that the descriptions of men's and women's online preferences relate to gender expectations rather than actual activities.

One 'Internet expert' confidently claims that 'everyone' who wants to use the Internet today can do so, and gender is no longer the most important difference. Several articles follow this pattern: a journalist trying to sketch a picture of gender as the main difference, following the previous hegemonic discourse, is contradicted by an 'expert' pointing out that the situation has changed. One expert agrees that there are some differences between girls' and boys' interests, but 'when entering working life, the computer knowledge of both genders is roughly the same.'[84] Another expert claims that it is primarily 'myths' about gender that create expectations of girls' and women's lack of skills: 'In reality, boys and girls have equal levels of knowledge.'[85]

Another example of old gender gaps being closed comes from a researcher commenting upon her own study showing that girls' and boys' use of the computer is converging. 'More boys are chatting...and more girls are playing PC games.'[86] But there are still considerable differences 'both in what they do and how much time they use'. She continues: 'I believe that it will always be like that, because it is based on fundamental differences in interests between the genders.' Despite the closing of old gaps, new ones appear, indicating the importance of gender in sorting the world, and also how easily this sorting mechanism can appear as a documentation of 'nature', an unchangeable and rigid structure (Connell, 2002, p. 68).

These articles illustrate how the hegemonic discourse is no longer in harmony with the observable practices of people. The hegemonic discourse is continually being renegotiated in ways that seek to challenge the masculine norm, and to revalue girls' and women's relationship to computers in more positive terms. These articles also remind us that gender is not a purely biological feature, but involves categories that include meaning, values and expectations. While old gender gaps might be closing, new ones appear and re-install gender as a main differentiating feature.

The fourth and final main topic since 2000 is the ongoing focus on girls' and women's low level of participation in computer education and the computing workforce. Here, the image of 'nerdy men' is still repeated (although not in such a celebratory tone), and we find echoes of the feminized discourse emphasizing computing as mostly about people and communication, which is assumed to interest women more than technical aspects. Women are also wanted in the IT business because 'they think a little differently and supply new perspectives'.[87] But the most interesting development in this discourse is new images of *men* fighting the image of the nerd/hacker/geek, which has been seen as one of the main problems for women in relation to computing, as representing what women are not (Turkle, 1988; Rasmussen and Håpnes, 1991; Lagesen, 2003; Hayes, 2010b). These new men are not isolated or asocial; neither do they have a 'single-minded devotion to computers' (Turkle, 1988, p. 47). Quite the contrary, they are active in various social activities and organizations, and 'not engaged with the computer all the time'.[88] The new men 'talk about other things than their computer' and even 'have a tan in the summer'.[89] The invitation to women is still not based on women being technically competent, but rather reflects a truism about women in professional contexts representing 'something special' purely based on gender. Similar perceptions of women representing something special have also been found in other Western countries, with expectations of women's social abilities (Woodfield, 2002, p. 121), sometimes resulting in women being 'pushed' towards soft areas with fewer career prospects, according to Barbara Bagilhole et al. (2008), or towards management, according to Webster (2005). It is, however, still not clear whether women really contribute something special (Sørensen, 1992; Faulkner, 2001; Mack, 2001), or whether they are socialized into the existing culture, as suggested by Abigail Powell et al., who claim that 'women engineers either share the values and attitudes of their male colleagues or that on embarking on an engineering career, women assimilate to the engineering culture, failing to challenge the dominant masculine discourse' (2006, p. 696). Most notably, what is being reconstructed in the examples given is not so much women or computer education – but rather men. Even so, hackers or 'computer idiots' still exist and 'we need them'.[90]

A number of articles focus specifically on the female leaders who make up around 4 to 5 per cent of the IT industry leaders in Norway. Female leaders are sought for their 'female competence' and to secure the 'needs of half the users',[91] even if both of these observations remained rather vague and unspecified (cf. Webster, 1996, p. 178). Women, however, 'don't know the code' of the IT business, and in recognizing this, a mentoring project initiated by a network for women in IT identified male 'agents' as mentors for female 'talents'. One of the mentors could report that his (female) 'talent' had become 'more rational and less emotional'.[92] Thus, there is a desire to increase the number of women in leading positions, to create role models and to help them learn the codes. This assistance is expressed by very rigid gender constructions, however, where women are marked as different from the male norm. Women can never win the competition with men's culturally acquired professional competence based on their 'nature' as women (Woodfield, 2000; Sørensen 2002b). And paradoxically, women are invited for some special values or skills they have as women, but to achieve the top positions they have to become less like women and more like men, for instance, in learning the codes of the (male) business. Thus, again we see how women meet double standards in computing. We saw earlier that when facing the masculine culture of computing, women who perceive computing to be gender-inauthentic might attempt to downplay their technical skills at the same time as social and communicative skills, mostly associated with women, might be valued more highly when found in men (Woodfield, 2002, p. 127). The act of balancing between femininity and technical skills can make women downplay their femininity or attempt 'to become more like men' (Bagilhole et al., 2008, p. 20). Adam et al. illustrate this with one of their informants, Vanessa, who 'downplayed her femininity, becoming a genderless "it" in an attempt to make herself less visible' as woman in an all-male environment (2005, p. 289). When she later decided to start dressing in a feminine way, she was mistaken for being part of the administration. Another woman interviewed by Adam et al. 'feels she is not meant to be technical because she is a woman whilst she cannot be a woman if she presents herself as technical' (2005, p. 290), clearly illustrating the conflict that women might feel in a masculine technological world.

The presentation above has focused on the new discursive constructions, although, it should be emphasized, the press at the moment also continues to write about girls and women that have less interest, less experience and less computer skill than boys and men. The most pronounced continuities concern the different computer users and the persisting low proportion of women in computing education and work. These are also the fields where gender seems to be most stubbornly reproduced, often in an essentializing manner, making it appear as if it 'will always be like that' owing to men's and women's 'nature' (Connell, 2002, p. 3). We have also seen new discursive constructions, in which discursive continuities operate simultaneously with new voices. The new voices introduce new and more varied groups of computer users. They try to restore some kind of gender balance by claiming things have changed and by introducing the new social men (with a tan) who are assumed to make the IT world more attractive to women. The expansion and diversification in these accounts make women less evident as *the* non-users of computers, even as the young teenage boys have lost some of their former status as computer wizards, softening the gendered pattern of visibility and invisibility of skills.

How can we understand the changing patterns?

In this chapter we have seen how perceptions of the relationship between gender and computer technology developed in popular discourse in Norway, and we have seen similarities to other Western countries. Thus, overall, the findings are perhaps not surprising, since the computer's masculine associations and the low proportion of women in computing are well-known (Turkle, 1988; Margolis and Fisher, 2002; Lie, 2003a; Cohoon and Aspray, 2006b; Misa, 2010b). This chapter's focus on discourse has, however, illustrated how perceptions of gender-technology relations have changed since the early 1980s. We have seen that the connection between masculinity and computers grew stronger through the 1990s, as well as indications of new images of computer users most recently.

During the period since 1980, we can see three main phases in the cultural appropriation of computers. First, in the early 1980s, before the dominant discourse took hold, the perceptions of gender were unclear and ambivalent, and descriptions of computer technology, its uses and (potential) users could even be distinctly hospitable to women. This was even more explicit in the Norwegian computer magazine *Datatid*. The magazine frequently suggested that the computer offered special opportunities to women by creating new jobs in or near the home for women with care-giving responsibilities (Corneliussen, 2006). Similar images of women as teleworkers also appeared in other parts of Europe in this period (Huws, 1991). This phase soon gave way to a flood of popular-culture accounts where computers and computing were clearly gender-typed as masculine. Men were increasingly perceived as active users while women were perceived as problematic non-users. The explicit gender-typing as well as the pattern of visibility and invisibility were major driving forces in establishing the hegemonic discourse of the late 1990s. The third period, since 2000, was more expansive; the discursive changes were closely tied to the spread of computers throughout society, and access to a PC in the home increased from 71 per cent in 2000 to 79 per cent in 2004 and 92 per cent in 2009, while access to the Internet grew over the same period from 52 per cent to 66 per cent and 91 per cent (Vaage, 2010, p. 80). Computers and the Internet had clearly reached a majority, and on the way contributed to expanding the image of computer users in ways that challenged the previous pattern of visibility and invisibility. The invasion of 'everybody' blurs the gender-typing and actively changes ideas about who are and who are not computer users.

These historical patterns raise at least three questions. First, why in the first phase did the computer appear to be ambivalent or even gender-neutral? Even though computers were not entirely new in the 1980s, computers in the home and in the office were new to most people. In the early 1980s, office workers perceived the computer as novel and exciting, but their excitement disappeared within just a few years (Lie, 1998). Also, early on, the computer was symbolically flexible enough to be associated with either the female office worker or the male office leader. Later, in the mid-1990s, Norwegian girls expressed a similar ambivalence about the Internet, making the researchers conclude that the Internet had not (yet) achieved the same strong associations with masculinity as the computer (Håpnes and Rasmussen, 2003).

Although it has been claimed that young girls can 'sense' the masculine character of computers, disposing females to reject them (Nissen, 1996, p. 153), what we have seen here indicates quite the opposite: that computers did not enter culture with a ready-made masculine symbolism attached. Instead, personal computers began with an unclear, ambivalent and somewhat confusing gender-typing. It is worth speculating why this might be so. While technology in general had connections to the heavy and noisy mechanical technology so deeply rooted in a working-class masculinity (Lie, 1998), the personal computer had other 'less straightforwardly masculine' qualities (Mahoney, 2001, p. 171), and optimists expected women to play a more important role in relation to computer technology. The new office computers were also linked to the secretary's typewriter (Wajcman, 1991, p. 150), and we recall how Judy Clapp's story illustrates how computer keyboards presented a challenge to male executives accustomed to dictating letters rather than typing them (see Chapter 1).

Second, why and how did the computer *become* clearly masculine through the 1980s and 1990s? Gender was used not merely to mark

differences between men and women, but also as an ordering structure that made men's and women's actions visible and invisible in a
certain pattern. Because of this ordering structure, what was made discursively visible is not necessarily the 'true' story (recall the invisible
legions of male non-users), but rather a partial story filtered through
gendered cultural expectations. When men – or rather a distinct minority of computer-savvy men – were taken to represent the norm of active
computer users, women were seen as deviants while men not conforming to the norm became invisible. Caroline Clarke Hayes suggests that
this image grew partly as a consequence of computer science becoming more popular, parallel with a growing 'general public awareness
of stereotypes depicting computer scientists as male "computer nerds"
and "hackers"' (Hayes, 2010a, p. 26). The obsessive focus on gender as
the main difference, even when reporting about a closing gender gap,
suggests that gender is the fundamental category we use to structure
our perception of society (Scott, 1988) and, accordingly, our expectations about computers. This does not only count for popular images in
the media (Kitzinger et al., 2008); Adam et al. present a similar criticism of research on gender and information systems. In particular,
quantitative studies where 'gender is seen as a dichotomous variable',
they argue, 'forces gender into polarised masculine and feminine categories therefore emphasising differences between the two' with the
consequence of 'making male and female appear as autonomous categories, de-emphasizing the effects of age, class and ethnicity' (Adam
et al., 2004, p. 229). The gender differences on which there is persistent
focus 'depend on the refusal or repression of alternative possibilities',
Scott points out (1988, p. 43), which in this case makes it difficult to see
commonalities and similarities in men's and women's experiences with
computers (Connell 2002, p. 42). Or alternatively, similarities between
the male and female non-users!

Third, why do we find a more marked discursive variation after 2000
that indicates change in the dominant discourse? The gender ambivalence of the early 1980s might be ascribed to the newness of the computer, a difference in kind. Since 2000, there has been an important
difference in degree, and 'more is different' (Shirky, 2008, p. 149).
We have seen that access to computers and the Internet grew rapidly,
exceeding 50 per cent in 1997 and 2000, and with 92 per cent of the
population having access to a computer and 91 per cent to the Internet
in 2009 (Vaage, 2010, p. 80).[93] Now, the main difference in access and
use is no longer between men and women, but between young and old
(98–100 per cent of those aged 45 and younger have access to a home

computer compared with 60 per cent of those aged 67 and older). It seems new technology becomes most interesting when it becomes trivial and available for everyone to use (Shirky, 2008, p. 105). Despite the massive growth during the 1980s and 1990s, it is primarily in the period after 2000 that computers and the Internet reach this phase of pervasiveness. Thus, the image of the user changes from young male enthusiasts to 'everyone' using it for 'everything'. Not only are more people using computer technology, but also there is a new way of perceiving computer technology. '[T]he generic concept of ICT is less meaningful to young people. They prefer to talk about specific activities that they perform using ICT...The issue is no longer whether or not to use ICT, but what activities you need ICT to do' (Gansmo et al., 2003b, p. 137). In the 1990s, 'access' was a key word in discussions about gender and ICTs, with an underlying assumption that 'access to the technology and information about its brilliance will make the women "change side"' (Gansmo, 2004, p. 87). Increased access and use has not so far resulted in more women aiming for a career in IT occupations. However, the developments since 2000 – the 'difference in degree' and trivialization of the computer – might also be important for the realization of this.

In this chapter we have seen not only spurs to change and variation, but also how discourses can suppress variation. We have seen something resembling a Bakhtinian choir, illustrating how discourses, even when they appear to be stable, should be seen as temporary fixed structures that will always be in motion, moved – although slowly – by our choices, statements, interpretations, actions and research. What we have seen here is that detailed analysis of public debates about computer technology might reveal hidden traces of gender in the history of computers. It can help to bring back to life stories about computer technology representing something special for women, as well as stories about male computer-reticence, both of which can provide a challenge to stereotypical images of men's and women's gender-technology relations.

'Don't call us retards one more time!!!!!!!!!!!!!!!!!!'

There is a strong sense of gender equality in Norway. We talk as if we (almost) have gender equality, and mentioning gender as a difference is not always considered appropriate (cf. Corneliussen, 2003a). However, when we talk about computers and technology, this does not seem to apply. Rather, a general acceptance of the computer as a boy's toy was established as common knowledge, and the claim about gender making a difference in relation to computers was not considered 'politically incorrect', as most other claims about gender making a difference in terms of skills or knowledge would be. And these claims did make a difference, as illustrated by a reader's letter from a girl aged 12 after the publicity of a survey showing that Norwegian girls used the computer less than boys:[94]

> I get really furious, or rather bloody furious when it is written that Norsk Gallup sort of has made an investigation about Norwegian girls as PC retards! Because it is not true! Where I live there are lots of girls who use the PC regularly every day! I also use it every day, for many hours. ... I just want to say that I am a girl, I know something, and I am not a retard!! You got the wrong impression of us girls. ... Don't call us retards one more time!!!!!!!!!!!!!!!!!!![95]

This quote is illustrative of how the discursive claim about girls' and women's low interest in computers could affect real people, and we see how this girl apparently felt that she had to use a loud voice to get her counter-discursive message across. What she reacted to was not the fact that girls in general used the computer less than boys but that girls were discursively constructed as one large group of 'PC retards', and to the exclusive focus on girls as less interested, ignoring the fact that 27.5 per cent of the girls in the survey *did* use the computer for several hours *every day*.[96]

Religious societies are excused from the law of gender equality in Norway (Skjeie and Teigen, 2003, p. 20). Computers are not. But while religious societies *practising* this excuse have in several instances met heavy criticism, claims of gender as a difference in relation to computers rarely do. The law of gender equality might guarantee equal rights, but cultural discourses are not easily ruled by national laws – discourses have their own laws, and they have their own power.

3
Discursive Developments within Computer Education

Introduction

Differences between men's and women's relations with ICTs received considerable attention during the closing decades of the twentieth century, both within research and in educational institutions. We have already seen the growing concern that girls and women would become the losers as non-users in a society where the importance of ICTs was increasing. However, shortly after the turn of the millennium, the gender gap in both computer access and use was about to disappear in the younger age groups. Since 2000 the gender gap seems to be better described as a combination of gender and generation (Gansmo et al., 2003a). Within computer education, however, the gender gap does not seem to want to release its grip. As we saw in Chapter 1, even recent gender and ICTs research points to how 'almost thirty years of efforts have failed to produce a sustained increase in women's participation in computing', and 'the situation is not likely to improve any time soon' (Cohoon and Aspray, 2006a, p. 139), and this is so even in Norway (Wyatt, 2008a). Although we can find the stability argument in relation to other contexts – for instance, about computer use in general – computer education has probably generated most claims about lack of improvement for women, making this one of the most important fields to explore in our search for change and stability in gender-technology relations.

By 1991 the overall proportion of female students in higher education had reached 50 per cent in Norway, and this has increased to about 60 per cent.[1] A similar 'feminisation of the student population' is found across Europe with an average of 55 per cent female students and 59 per cent graduates (European Commission, 2009, pp. 7, 66) and in the US.[2]

Thus, we know that women do not object to studying. While other fields experienced a steady increase in the number of female students, this has not been true for a handful of subjects that are still strongly male-dominated, such as computer science. This means that there is a flaw in the 'critical mass' argument predicting that once a greater number of women enter a specific educational field, it will make it easier to attract more women – a discussion to which we will return in Chapter 7. The low proportion of women in computer science or informatics clearly gives a gloomy picture, and Cohoon and Aspray are correct in pointing out that the efforts to increase women's participation have largely failed, both in the US (2006a) and in Europe. Charles and Bradley found an under-representation of women in computer science in the 21 industrial countries they studied, although to varying degrees. The situation was less severe in the former Soviet states, where Charles and Bradley suggest that the educational policy of sorting students 'based strictly on academic performance' might have 'reduced the influence of essentialist preferences and stereotypes in the educational sorting process' (2006, p. 191). Norway, together with the UK, Australia and New Zealand, was placed in the central band in Charles and Bradley's study. Thus, despite the strong position Norway has in terms of gender-equality measures, the number of women in computer science has been low since computer science or informatics became a separate subject at universities in the late 1970s. A rapid increase in the 1980s was followed by a decrease towards the mid-1990s, a tendency also observed in other Western countries (Stuedahl and Braa, 1997; O'Leary, 1999; Woodfield, 2000; Mack, 2001; Camp and Gürer, 2002; Barker and Aspray, 2006; Cohoon and Aspray, 2006a). The decrease in the number of women in computing in the mid-1990s was followed by a number of initiatives to recruit and retain women, most of which had positive effects, although only temporary. Following the turn of the millennium, we have again seen a decrease in the number of women in the already mostly male-dominated areas of computer education.[3] Hayes points to a similar development in computer science in the US, with a rapid growth between 1972 and 1984 and the subsequent decline. The proportion of women fell by 17 per cent between 1984 and 2006, thus making computer science appear to be 'a field of extremes', both in growth and decline, compared to other educational fields (2010a, pp. 31–2).

The lack of improvement in the proportion of women within computing does not, however, mean that nothing has changed in computing, and in this chapter we will explore the discursive development within computer education in Norway – more specifically, how the perception of gender and computer education has changed since the early 1980s. The recurring questions have been: Why are there so few women in computer education? How can this be changed? Thus, discussions on gender and computer education have mainly focused on women, and on how to recruit and retain women in informatics and computer science (Faulkner and Lie, 2007; Schiebinger, 2008b; Trauth et al., 2009). The arguments as to why women should be recruited to ICT education have been varied. While gender-equality in education and in the job market has been important to some, others have claimed that computer technology shapes society, thus arguing that women should also take part in the development of technology. While some have suggested that women have something special, feminine values, that the ICTs business needs (Teigen, 2000; Camp and Gürer, 2002; Peters et al., 2002; Sørensen, 2002b), others have emphasized that the motivation to reach gender-equality should not be based on assumptions about gender qualities, but simply on a notion of justice and balance within important institutions and fields in society (Skjeie and Teigen, 2003, p. 28). The need for labour within the ICTs business is a recurring argument (Ahuja, 2002). More recently, this has been followed by an increasing emphasis on IT business needs, which include 'a wide variety of "types of people"' (Faulkner, 2006, p. 10) as well as the risks incurred if that is not achieved. These risks are summed up in the report *Women in IT: The Facts* as threefold; first, 'a shortage of talent' and thus 'a devastating loss of capital'; second, 'reduced innovation, productivity and competitiveness'; third, 'financial losses and decreased customer satisfaction' (Ashcraft and Blithe, 2010, p. 10).

We will leave the arguments for why women should be recruited here, and focus instead on different strategies with which the low proportion of women in ICT education has been met in Norway since the early 1980s. We will see how various inclusion strategies have been based on different discursive perceptions of the relationship between gender and ICTs, and consequently, how they seem to justify different strategies to include more women.

The last decades of feminist technology research have supposed that technology is not neutral in relation to gender (Cockburn, 1992). We have learnt that the relation between gender and ICTs is not predetermined by some internal essence, but rather is shaped within a social and cultural context (Wajcman, 1991; Lie and Sørensen, 1996b; Cockburn and Ormrod, 1993; Lie, 2003a), as well as by the individual's situation (Trauth and Howcroft, 2006). Thus, the discourse of ICTs can be seen as a structure of meaning in constant movement: the hegemonic discourse,

as we have seen in Chapter 2, is not fixed, but is constantly being challenged and influenced by potentially new meaning, by non-hegemonic
articulations. The discourse can be negotiated, and in the process, perceptions of gender and technology are intertwined and woven together.
In this chapter we will see this illustrated by three different discursive
constructions of gender and ICTs that have appeared in debates since
the early 1980s on how to include women in computer education in
Norway. We will also see how the different discourses create different
foundations for how to address the low proportion of women in this
field. The first is a gender-blind discourse in which gender is not recognized as an important category. The second is a masculine-connoted
discourse, firmly rooted in a notion of the close connection between
masculinity and computers. The third is a *feminized* discourse. This
does not represent a feminine opposite to the masculine discourse, but
rather a *feminine turn* of the masculine discourse, a turn that introduces
femininity without challenging the close affinity between masculinity
and technology. There is a certain chronology in the discourses presented here, but they should not be seen as stages that succeed and take
over from each other., Neither are they mutually exclusive; they can –
and indeed do – exist side by side, even within the same institutions. It
should also be noted that this is by no means an exhaustive documentation of gender politics within computer education in Norway, but is
a selection of gender political actions that will serve to illustrate different discourses and different ways of building inclusion strategies. By
exploring different strategies to recruit and retain women in computer
education, we will be able to see how different discourses create different foundations and premises for the efforts to bridge the gender gap in
computing. As we will see, all three discourses have their weaknesses,
and the final question we will pursue in this chapter is how we can
challenge the problematic aspects of these discourses.

The gender-blind discourse of computer technology

At the end of the 1970s, feminist researchers took up the debate on
women's marginal position within natural sciences and technology.
Science and technology, as well as related skills, artefacts and cultures, represented men's interests rather than women's interests, it was
claimed (Woodfield, 2000, p. 1). In contrast to traditional 'noisy' and
'dirty' mechanical technology, the clean and quiet information technology represented something new, with other qualities that were just
as relevant to women's nimble fingers (Wajcman, 1991, p. 150) as to

male technological traditions (Mellström, 2004; Ensmenger, 2010). The new computer technology did not, however, work as a symbol for working-class masculinity in the same way that mechanical technology did, according to Lie. The computer could instead act as a symbol for other kinds of masculinities, based on intellect, control and leadership. It could even work as a symbol for the female secretary's working tasks (Lie, 1998), although, not primarily as 'technology', but rather as *office equipment*. The computer did, in other words, contribute to a disturbance in the traditional image of men's close relationship with technology. Optimistic as well as pessimistic visions of the future have followed the advent of the computer (Faulkner, 2001; Wajcman, 2004; Wyatt, 2008a), from a hope that women would not be marginalized in relation to computers (Ahuja, 2002), to a fear of technology as a threat to women's jobs (Lie et al., 1984; Barker and Downing, 1985 [1980]; Webster, 1996; Lie, 1998; Woodfield, 2000; Gansmo et al., 2003a). The most negative predictions have not come true, but neither have the most optimistic visions; women did become a marginal group in relation to the new information technology (Rasmussen, 1988, p. 74; Sørensen, 2002b, p. 23). During the 1980s, the lack of women within informatics was increasingly being scrutinized, at first from within, by female information scientists. Information scientists Tone Bratteteig and Guri Verne wrote in the Norwegian journal for women's research (*Nytt om kvinneforskning*) in 1985: 'How can we apply a female perspective on the discipline of informatics?' (p. 9). They pointed to a number of areas within informatics that they believed could be seen from a female perspective. Simultaneously, they pointed out their own limited background within women's research and theories, and encouraged other women's researchers to contribute their knowledge to the debate.

Initially, feminist criticism of technology was inspired by criticism of natural science's apparently objective production of knowledge (Harding, 1986; Haraway, 1991c, 1997). Feminist researchers have pointed out how the connection between technology and natural sciences has given technology a similar image of being a given object, that is, out of reach of what is socially constructed, and unaffected by social factors, such as, for instance, gender (Bratteteig and Verne, 1985; Harding, 1986; Asdal and Myklebust, 1999). Disproving this assumed gender neutrality, feminist researchers have illustrated how information technology involves gender in a number of different ways and on several levels (Harding, 1986; Mörtberg, 1994; Faulkner, 2000a; Bagilhole et al., 2008). Questions concerning gender and information technology were not new in 1985, but Bratteteig and Verne's article indicates that these questions had not yet been put on the agenda within informatics and computer science. And when gender is not perceived as an important issue, it often remains invisible. In subsequent decades, a number of researchers have illustrated the close connection between masculinity and information technology (Sørensen and Lie, 1988; Hacker, 1989; Wajcman, 1991; Lie and Sørensen, 1996b; Lie, 1998; Margolis and Fisher, 2002; Lie, 2003a; Cohoon and Aspray, 2006b; Misa, 2010b). Explanations of the bond between men and technology have varied, for instance, by Knut H. Sørensen and Merete Lie pointing to the traditional division of labour between men and women, claiming that technology used by women has not been perceived as 'real' technology (1988). Information scientist Cristina Mörtberg emphasizes the male dominance within fields that traditionally have been important in the development of ICTs, as well as observing that career patterns in computing are unfavourable for women (1987, 1994). Social scientist and psychologist Sherry Turkle gives an object relational explanation to girls' and boys' different relationships with the computer (1984), and social scientist Jörgen Nissen claims that girls can sense the male-dominated history of the computer (1996). The close connection between men and technology, which runs parallel with the invisibility of women's relations to technology, has resulted in technology rubbing off on masculinity in ways that are difficult to separate from masculinity rubbing off on technology (Sørensen and Lie, 1988, p. 11). In more recent research, negative images and stereotypes emphasizing computing as a male domain have been presented as one of the main reasons for women not choosing computing (Cheryan et al., 2009; Hayes, 2010a). Young women do not want to be perceived as masculine (Mack, 2001), and 'girls are anxious to adapt to the female image' (Sagebiel and Dahmen, 2006, p. 11), which is not in harmony with computing. 'The door may be open, but the world beyond it does not invite entry' for women, Mahoney points out (2001, p. 171).

As we know from a number of contexts, when women are not visible, gender is rarely perceived as an important issue, which also has been the case with technology. However, the fact that gender is not being explicitly discussed does not mean that it has no relevance. Bratteteig and Verne claimed that the close connection between men and informatics made women feel like strangers in this field. Male dominance made women particularly visible in the environment, and they were often perceived as 'representatives for the female gender' rather than specialists in computing (1985, p. 14), similar to what has been reported in other Western countries and also in more recent research (Salminen-Karlsson, 1999; Adam et al., 2005).

Several research projects from the 1980s have also demonstrated that women felt marginalized, both in the male-dominated environment of computing and in relation to the professional norm (Kvande, 1984; Håpnes and Rasmussen, 1990; Rasmussen and Håpnes, 1991; Håpnes, 1992). Women preferred subjects within ICT education that have been described as 'soft', where the use and usefulness of ICTs are central. In a study from a Norwegian technical college, Tove Håpnes and Bente Rasmussen found that a minority group among the students – the hackers, or 'key pressers', as the female students called them – were perceived as a symbolically important group. The hackers, which were exclusively made up of young men, were more engaged in topics that were perceived as 'hard' topics within informatics, such as programming and hardware-related subjects. The female students felt that the 'hard' topics were perceived as more important topics than the 'soft' topics they had chosen (Håpnes, 1992, p. 161). The hackers or young men obsessed by computers have been found to be a problematic group for women in other countries (Turkle, 1988; Margolis and Fisher, 2002), as has a similar association of men and women regarding different topics within computing and engineering (Faulkner, 2006; Phipps, 2007). Sub-sections of science, technology and engineering have been feminized, Bagilhole et al. suggest – however, with the effect that women find themselves in 'female ghettoes' in an otherwise masculine world (McIlwee and Robinson, 1992 cited in Bagilhole et al., 2008, p. 8).

The early period of informatics and computer science in Norway was one without specific strategies that aimed at changing gender inequalities within educational institutions. But neither were there any active strategies to prevent women from entering computer education; women did have access. Or, as one of the employees at the institution studied by Håpnes and Rasmussen expressed it: 'They [women] can [study computer science] as long as they want to and have the courage to do so' (Håpnes, 1992, p. 173). Lack of strategies to deal with the gender question can perhaps be comprehensible if we see this in a perspective that combines gender-equality and modernity (Giddens, 1991; Annfelt, 1999), emphasizing, first, that men and women should have the same opportunities, and second, that the modern woman is responsible for achieving her own true potential. However, as Elin Kvande and Bente Rasmussen point out, equal treatment of men and women might result in inequality when men and women are starting from different premises, in which case equal treatment becomes a 'hidden gender ordering process' (Kvande and Rasmussen 1993 [1990], p. 123, cf. Rees, 2001; Bagilhole et al., 2008; Hayes, 2010a). Those hidden processes are hard to get to grips with, because they give the impression of practising equality. And as we will see in the following chapters, men and women do indeed have different premises to start from in computing.

The way of treating gender in ICT education seen above exemplifies what can be described as a gender-blind discourse, where the connection between gender and computer technology is not explicitly recognized. The fact that more men than women choose to study information technology is not seen as a result of either technology or education having anything to do with gender, but rather as the result of men showing greater interest in computers. The perception of technology as gender-neutral makes it possible to conclude that the problem is instead a question about will and courage among women. A gender-blind discourse offers a poor foundation for gender-political actions. For educational institutions, the gender-blind discourse meant that they could (continue to) relate to the professional tradition developed in a male-dominated environment without posing critical questions involving gender. Feminist researchers and female information scientists, however, did raise questions about gender on the international scene with strong contributions from, for instance, Cynthia Cockburn with *Brothers: Male Dominance and Technological Change* (1983), Sandra Harding with *The Science Question in Feminism* (1986), and Judy Wajcman with *Feminism Confronts Technology* (1991). In Norway, a group of researchers from Trondheim explored gender and technology, and asked 'Does technology have anything to do with women?' Primarily interested in the introduction of technology into working life, the researchers explored gender and technology at work in a number of different contexts. They illustrated how men were working closer to technology and also dominated within the production of technology, thus emphasizing the gendered division of work as one of the main reasons why technology had become a male domain (Lie et al., 1988).

According to the psychologist Hanne Haavind, male dominance has to be made visible before we can fight it (1986), and it was the task of making gender visible within ICT education that was of primary importance in this early period. There were a few attempts at making women choose 'untraditional career patterns' and a few examples of using 'soft gender quotas', but these had a minor influence on gender distribution at higher levels of technological educational institutions (Kvande, 1982 cited in Kvande and Rasmussen, 1993 [1990]; see also Verne, 1988; Teigen, 2000). Several years passed before more extensive

measures to increase the proportion of women within computer education were initiated, and rather than this being a direct result of either
internal or external criticism from gender researchers, it was an answer
to a situation where the number of female students had decreased even
further during the first half of the 1990s (Stuedahl, 1997a; cf. Hayes,
2010a).

The masculine discourse of computer technology

The decreasing number of female students was perceived as a crisis at
several of the higher ICT educational institutions in Norway. The character of the crisis at the Department of Informatics at University of Oslo
was primarily that progressively fewer women completed the educational programme. The number of female students studying lower-level
degrees was more or less constant, while the number of female master's
students declined from about 20 per cent in 1985 to 8 per cent in 1995.
A project called 'Girls and IT' was initiated to deal with this crisis, and
this project was mostly targeted at the 'shrinking pipeline' between
bachelor's and master's degrees, a challenge also found in the US (Camp
and Gürer, 2002).[4] An important objective for the 'Girls and IT' project
was to encourage female students who had already entered IT education to continue and complete their education at master's level. This
project illustrates a new way of dealing with gender within ICT education based on a masculine discourse, explicitly rooted in a masculine
norm in relation to information technology. Initially, a research project
was undertaken in order to build knowledge of the situation (Stuedahl,
1997a, 1999; Langsether, 2001). Male and female students at lower levels
were interviewed in order to learn more and to reach an understanding
of 'girls' experiences of the everyday life at the department' (Stuedahl,
1997b). Dagny Stuedahl found a number of reasons why women found
the situation at the department difficult. Some of the female students
found the study difficult, some felt marginalized as a result of male
students' knowledge and extensive use of computer terminology, and
the women were not satisfied with the male-dominated social environments (Stuedahl, 1997a). The findings of this study resulted in several
remedial actions, of which we will explore two that can be characterized as immediate damage control measures aimed at controlling the
declining number of female master's degree students.

One of these measures confronted the gender distribution of the
students. The low number of female students made it likely that there
might only be one woman in a student group. To counteract this it

was decided that four out of a total of 20 student groups at lower levels should consist of an equal number of male and female students. The other damage control measure focused on women's lack of knowledge, and female students were invited to a special course to learn 'the things we assumed that the boys knew about computers before they attended the program'.[5] The student groups with equal gender balance were unofficially labelled 'girls' groups' (Langsether, 2001, p. 8), while the course offered to female students was referred to as a 'boy's room competence course' by the employees at the department (Braa, 1996; Stuedahl, 1997a).

These measures were based on a perception of women as representing a minority, both socially and in relation to knowledge. The measures were twofold: they targeted the challenges of creating a female community where the number of female students was low, and they tried to ease women's experiences of male students being more knowledgeable, and consequently the difficulties women met with participating in a social environment where male students created the 'premises for social interaction' (Stuedahl, 1997a; Langsether, 2001, pp. 84–5, cf. Björkman, 2002, pp. 61–2; Sørensen, 2002b, pp. 18–19). Jane Margolis and Allan Fisher documented a similar tendency at Carnegie Mellon University, where female students entered computer education with a genuine interest in the study; however, when meeting male students who had more knowledge than themselves, they redefined themselves as being not particularly interested. This resulted in quite a few of the female students leaving the course, giving the reason that 'they were not interested after all'. One solution to this at Carnegie Mellon University was to offer differential points of entry to the study based on individual knowledge (Margolis and Fisher, 2002). At the University of Oslo, the girls' groups were presented as a means of giving women the opportunity to socialize with more 'like-minded' students, as well as giving them a learning environment among students with a similar level of knowledge. In other words, there were attempts to help, and more or less protect, the female students in the highly male-dominated environment.

The boy's room competence course, however, aimed at giving female students the same computer skills that it was assumed the male students already possessed. All over the Western world, informatics has been associated with a special kind of geeky masculinity, according to Vivian Lagesen (2003). The geek or nerd is the image of a boy who has a special and strong interest in computers, and learns a great deal on his own at home in the 'boy's room'. And it is this stereotype that is the origin of the concept of 'boy's room competence'. At the 'boy's room competence course' it was the *extracurricular* computer knowledge associated with men that was claimed to be important.

In contrast to the gender-blind discourse, the construction of gender in relation to information technology is explicitly employed in these measures; information technology is constructed as belonging in the boy's room. Women are constructed as aliens, as 'the others', those without a natural position within ICT education, and the solution is to adapt women to the norm.

It is easy to point to the flaws in these measures, as they obviously construct a general divide between men and women. The 'boy's room competence course' contributes to the construction of a professional norm based on gender that actively excludes women. It is also necessary to question the way women in a professional setting are encouraged to adapt to a norm based on young boys' non-professional arena. The aim of the girls' groups was to provide female students with the ambience of a safe learning environment, a goal that was achieved (Langsether, 2001). However, this implied assistance to integrate women in the masculine norm. What kind of knowledge should 'count' or define the standard of informatics was not questioned. It is also pointed out by Powell et al. that '"fitting in" reinforces, rather than challenges, the dominance of the majority group' (2006, p. 691).

Men's relationship to information technology and the relationship between masculinity and ICTs has been explored in a number of studies (Turkle, 1984, 1996; Nissen, 1993; Aune, 1996; Håpnes, 1996; Mellström, 1996, 1999; Berner and Mellström, 1997; Lie, 1998; Corneliussen, 2003c), including this book (see Chapter 4). Technology researchers have argued that the knowledge associated with the boy's room is not only assumed to be an advantage for boys attending ICT education, but is also seen as important knowledge to possess in order to be perceived as a proper 'computer person' (Sørensen, 2002b, p. 24). Sørensen points out that boy's room competence is something that only a small number of boys have, contrary to the majority of boys and girls. Treating this rather diffuse knowledge as highly-valued computer knowledge creates a situation where being a boy becomes synonymous with having computer knowledge, while being a girl becomes synonymous with not having these skills (Sørensen, 2002a, p. 9). However, from their study at Carnegie Mellon University, Margolis and Fisher found that boy's room competence is not important in relation to more lengthy ICT studies (2002), which is in line with a large-scale study of professional programmers documenting that the likelihood of producing correct programming code increases substantially and proportionally with length of education.[6] Also, during the 2000s, the differences previously found in men and women's programming experience before attending higher education is changing, according to Hayes. Changing requirements for prior programming experience might therefore not only benefit women, but might also help to recruit a more diverse group of men to computing (Hayes, 2010b, p. 270). Thus, we should question the role that boy's room competence was allowed to play within formal computer education. And perhaps we should also question why this knowledge was described as 'extracurricular' if it was regarded as central to the study – questions that were given only superficial consideration (Stuedahl, 1997a, p. 46).

The girls' groups and the boy's room competence course were both regarded as successful. Women who participated in the girls' groups experienced them as positive, both because a greater number of women undertaking the study made the choice of being a woman studying informatics appear more 'normal', and because they felt more equal to the general level of knowledge in the group (Stuedahl, 1997a; Langsether, 2001, pp. 86–87). However, these initiatives also met with resistance among female students who feared that it would make women appear to be a special group in need of special treatment (Langsether, 2001, p. 91). As we have seen, both the girls' groups and the boy's room competence course can be characterized as assistance for women to enter a male domain. 'Women can as long as they want to and have the courage to do so' has become 'women can as long as they can fulfil the masculine norm'. Neither of the damage control measures confronts critical questions about the content of the study, only questions of how the content should be passed on to female students. Also, both measures are based on a static perception of gender, rooted in an assumption that men have a higher level of computer knowledge than women when they enter the study course. This creates two possible subject positions, a male subject position associated with computer knowledge and a female subject position associated with the lack of computer knowledge, simultaneously as a perception of what is considered the 'right kind' of computer knowledge is established.

The signal sent out by the boy's competence course is, according to Stuedahl, that 'it is not necessary with computer skills in order to study informatics, because you will get these at the department' (Stuedahl, 1997a, p. 46). But it also sends out signals that women – because they are women – do not fulfil important requirements for being a student in informatics. In other words, these measures seem to support and reinforce a widely recognized perception of women and information technology as opposites (Mörtberg, 1999; Powell et al., 2004; Adam et al., 2005; Phipps, 2007; Bagilhole et al., 2008).

There are also other equality measures in computer education that can be seen as influenced by this masculine discourse and as based on a strategy of helping women into a field that is discursively constructed as a male domain. Initiatives such as professional and social gatherings, extra and also extracurricular classes, female computer lab assistants, classes and computer labs for women only, and quotas for women are all measures that were tried out in the 1990s. However, initiatives treating women as a special group have also met with criticism. While some have feared that they could make women a 'secondary group', others have claimed that women should learn to handle male dominance in the educational setting because that is what they will meet as IT professionals (Stuedahl, 1997b, p. 16). In Adam et al.'s study of women in the UK IT labour market, one of the informants emphasizes this by claiming that IT is 'the ideal career for women who have grown up with lots of brothers' (2005, p. 295). Adam et al., however, warn that this perspective individualizes the problem by making the solution dependent on 'the ability of a woman to "cope" in an all-male, sometimes, sexist environment, rather than how this environment can be changed to make IT more amenable to all women' (2005, p. 293).

As mentioned, different discursive constructions can exist side by side, and the perception of gender and ICTs at the Department of Informatics at the University of Oslo was not limited to the discourse discussed so far. The next discursive construction we will explore was also echoed at the University of Oslo, illustrated by one of the lecturers writing in a newspaper: 'We do not want to leave the arena to young, technology fixated boys who are absorbed in making technical genius solutions to something that might be the "wrong" problem' (Braa, 1996). This articulation illustrates the change of direction of the third discursive construction, where 'boy's room competence' is no longer treated as a prerequisite for studying ICTs, but rather as less important knowledge. As we will see, this resulted in a feminized version of the masculine discourse.

The feminized discourse of computer technology

The third discursive construction of gender in computer education – the feminized discourse – will be illustrated by an advertising campaign launched by the 'Women and Computing' initiative at the Norwegian

University of Science and Technology (NTNU). The advertising campaign is only one of several initiatives launched by the 'Women and Computing' project. Other activities, such as professional and social gatherings for women, a woman-only computer lab, and a special quota for admitting women to ICT education were also part of this project. The first advertising campaign in 1997 aimed at recruiting more women to ICT education programmes, and the campaign has subsequently been repeated with slight variations several times.[7] The advertising campaign also received financial support from the Norwegian Research Council in 1998, to act on behalf of all the Norwegian universities and colleges offering computer education, thus it also represents a number of initiatives operating on a national level in Norway during this period.

The advertising campaigns were targeting young girls with the message that 'computer education needs girls'. The first campaign in 1997 was designed as a cinema commercial with a series of pictures to illustrate the 'need'. The first image presented an image of a boy as 'technical genius' and clever with computers, followed by an image of a girl as communicative, clever with people and analytical. The third image stated that 'the computer science programme is more about human beings than about machines' (Lagesen, 2003, pp. 80–81). In the online material of the campaign in 2002, it was claimed that 'Girls and boys are different and think in different ways. We have experienced that girls are more concerned with computer systems' usefulness and use, while boys tend to get absorbed by technique and details. ... Girls like to work with people; they are good at listening and talking together. Precisely these qualities are necessary in the development of computer systems in order to satisfy the users' needs.' A 'human perspective' on technology is emphasized, with a focus on the technology's ability to satisfy users' needs. This perspective makes it possible to establish a connection between girls and technological education, based on a perception of girls having special abilities to understand other people, to socialize and be cooperative (Woodfield, 2002). A 'human perspective' on technology, based on social and communicative skills, is constructed as synonymous with a female perspective (Lagesen, 2005), closely tied to femininity, as stated in the campaign by emphasizing that the 'department is ready to welcome you. Because it seeks what you've got: "feminity"' (Lagesen and Sørensen, 2009, p. 136). However, as we saw in Chapter 2, this involves a gender dichotomy in which 'social' and 'technical' appear as opposites and as 'mutually exclusive' (Faulkner, 2006, p. 5), which is a perception recognized as one reason for women being pushed towards the 'soft' areas of IT or engineering (Webster,

2005; Bagilhole et al., 2008, p. 31). The discursive reconstruction in the campaign also attempts to override an earlier reported 'incompatibility' between femininity and technology (Lockheed, 1985; Turkle, 1988; Wajcman, 2004), which has been recognized as a reason for women not wanting to enter computing and engineering (Mack, 2001).

A special space for women is created in this discourse of information technology, based on assumptions of female qualities. At the same time, male qualities are described as less valuable, illustrated by one of the involved participants at NTNU claiming that: 'The typical computer operator is male, a loner with an intense relationship with his computer, 24 hours a day. The social importance of the man and the computer's activities is vague' (Austgulen, 1996, quoted in Teigen, 2000, p. 136). It is no longer men but women who have the qualities that are needed in computer education. Thus, women are constructed as *better* than men in those aspects that computer science is *really* about (Lagesen, 2003; Lagesen and Sørensen, 2009).

The discourse of information technology is feminized primarily through demasculinization, by reducing the importance of the technical aspects of ICTs that are being associated with men and masculinity, according to Lagesen (2003). The redefinition of ICTs is, however, not followed by a similar redefinition of gender. On the contrary, Lagesen claims, this discursive construction is firmly rooted in a well-known dichotomy associating boys with technology and girls with social relations. Thus, the 'newness' of this discursive construction is not primarily the content, but the way in which the discourse is turning gender and technology upside-down, also illustrated by the headline on the campaign's website, stating 'Girls at the Top'. The new feminized discourse is constructed by *adding* female qualities, not by challenging the understanding of men as technically competent. Simply reversing 'the hierarchy of the social/technical binary' serves to strengthen the traditional binary, Lagesen and Sørensen point out (2009, p. 137). The perception of women in the campaign has roots within traditional gender stereotypes, and women are, among other things, assured that it is possible to combine being a computer scientist with being a mother. The redefinition of what was considered important within ICT education was, however, not accompanied by changes in the curriculum. Rather, the 'play on stereotypical images of women as a way to cater to the need for communication skills made it less necessary to change the curriculum', according to Lagesen and Sørensen (2009, p. 136). The discursive modification of the technical aspects through the campaign did, however, result in women entering the study with expectations that did not conform with the reality they met (Lagesen, 2003). The technical aspects were still important within computer science, despite the campaign's emphasis on the 'human perspective'.

Contrary to the promises of the importance of communication made in this campaign, Lagesen and Sørensen found communication skills to be more highly-regarded in software companies than in the educational context. While communication was described as being based on technical skills and as 'a hybrid, socio-technical activity' (p. 140) in software companies, 'only "technical" elements' (p. 137) were included in what was described as the 'core competence' of engineering education (Lagesen and Sørensen, 2009). Also, in her study of global computing organizations Woodfield has found that computer 'wizards', a label more likely to be applied to men than to women, were associated with the most 'productive' social and communicative skills (2002, p. 127). Thus, in the studies by Woodfield, as well as Lagesen and Sørensen, the social-technical binary appears to be less exclusive than the mutually exclusive dichotomy found by Faulkner in her study of engineers (Faulkner, 2000a, 2005; Lagesen and Sørensen, 2009).

Even though women are not described as particularly interested in information technology, they are described in relationship to the technology, which can be seen as discursively new compared to the masculine-connoted discourse. Several research projects have focused on how and for what girls and women have *used* technology since the mid-1990s (Håpnes and Rasmussen, 2003; Nordli, 2003). These have been able to document that girls and women do in fact use information technology, but often point out that they prefer other activities and have different motivations than men (Lockheed, 1985; AAUW, 2000; Becta, 2008). Håpnes and Rasmussen's study of teenage girls' computer use illustrates that girls use the computer for a number of different activities, and they describe different arenas for girls' and boys' computer use. 'The play corner' and 'the software workshop' are associated with boys, while 'the typing room', 'the information market' and 'the meeting-place' represent activities that girls enjoy (Håpnes and Rasmussen, 1999, 2003; cf. Nordli, 1998). Writing and searching for information are associated with more traditional female activities, while the Internet is a new arena, a new type of computer use that the girls in this study did not associate with the image of the asocial computer nerd (Håpnes and Rasmussen, 1999). In studies focusing on girls and women's computer use, it has often been claimed that they are motivated by usefulness and social arguments for using the technology (Lockheed, 1985; Valenduc et al., 2004; Becta, 2008). Thus, the discursive construction of the advertising campaign is supported by research documenting girls and women's special preferences related to information technology.

We saw that the masculine-connoted discourse was developed into an inclusion strategy that offered women *assistance* to enter the male dominated area of computing. The advertising campaign, however, develops a new inclusion strategy by creating a new discursive under-standing of ICT studies; a feminized discourse that is more in harmony with qualities that are already associated with women. This discourse is not an equivalent to the masculine-connoted discourse; the masculine discourse is not expunged, and the masculine connotations of techno-logical knowledge are still valid (although not as important as before). However, a feminization of the discourse has occurred, primarily by adding values associated with women; a new space for communicative qualities assumed also to coincide with women's strengths.

This way of constructing gender as a difference has met with criticism (Adam et al., 2004). Sørensen has pointed to how the technical com-petence associated with men is different from the social competence associated with women; while men's competence is associated with cul-ture, as a skill that has to be acquired, the social competence associ-ated with women is constructed as nature, as something women simply have (2002b, p. 25). The gender dichotomy has also been criticized in other ways, among others by Faulkner who illustrates how many of the dichotomies that contribute to an image of technology as gendered are not in line with practical work with technology where, most often, both sides of the dichotomy are present (Faulkner, 2000a, p. 760). Sherry Turkle, for instance, in her pioneering work on people's relations to the computer, illustrates the use of gender dichotomies when she describes a soft and a hard programming style, associated respectively with girls' and boys' ways of writing programming code (Turkle, 1984). Despite the gendered dichotomies not being in line with practical work on the computer, they still have discursive power (Faulkner, 2000a, see also Levold, 2002).

It has also been pointed out that a continuous focus on gender as a dichotomy will continue to construct women as 'the others', as oppo-sites to men, who most often represent the discursive norm (Adam et al., 2004; Phipps, 2007). Lagesen emphasizes the danger that the reversed dichotomy employed in the advertising campaign will snap back to its traditional position; the masculine above the feminine and the tech-nical above the social (2003). Woodfield illustrates this in a study of English computer companies; also, there the social aspect of computing is emphasized as particularly important in order to recruit women. However, in the actual job situations it was the technical aspects of the profession that were most highly valued (Woodfield, 2000). Women's qualities associated with nature are deemed to lose in competition with men's qualities associated with culture; with control, acquisition and intellect, according to Woodfield. Thus, the focus on gender differences in the advertising campaign seems to contribute to a dichotomized and essentialist perception of gender, as a dualism in which the two opposites can never become equals.

Were the inclusion strategies successful?

We have seen how the gender-blind, the masculine-connoted and the feminized discourses create different foundations for how to deal with gender within ICT education, and simultaneously, different perceptions of what ICT education is or should be. We have moved from 'women can as long as they want to', to 'women can as long as they can fulfil the masculine norm', and further, to 'women are best'. Were these attempts to include women successful? And if so, in relation to what? More women were recruited and more women chose to continue and complete their ICT studies. However, the computing educations discussed above are still male-dominated. The initiatives have also met both internal and external criticism (Stuedahl and Braa, 1997; Stuedahl, 1997b; Teigen, 2000; Björkman, 2002; Lagesen, 2003). Most severe perhaps is the criticism against the static and dualistic description of gender, as it becomes a source of limitation for both men and women. None of the discourses discussed in this chapter question the way *men* and not women are assumed to have technical competence. The initiatives we have explored did have some of the intended effects – for the women already participating and also a short-term ability to recruit women. The discursive constructions accompanying the initiatives are, however, not unproblematic. The static descriptions of gender we have seen in the gender-blind, the masculine-connoted and the feminized discourse make it difficult to create real or lasting change, indicating that we might need other discursive constructions in order to achieve that. Most important for such an alternative discourse would be to challenge the static and dualistic construction of gender-technology relations.

An alternative discourse could, for instance, include ideas allowing a wider spectrum of diversity in gender-technology relations. We could, for example, explore computer education itself. Even though publicity in the press during the 1980s and 1990s gives the impression that women in general are not interested in computer education (see Chapter 2), that does not give the complete picture. There are, as we will see in Chapter 4, computer education study courses with a high proportion of women – in some cases even more women than men. Thus, one question we could ask is whether we can learn something from those ICT studies that already have succeeded in attracting women. Second, feminist critiques have challenged the tradition of knowledge within computing, claiming that the tradition itself is a barrier to diversity. However, computing could be changed, for instance, by scrutinizing the knowledge tradition and its terminology that originated in a male-dominated environment, and by including perspectives that are seen as also attractive to women. Finally, feminist technology research has contributed knowledge that challenges the dualistic perception of gender and information technology through a large number of empirical studies, including the present work, demonstrating that the real world is too complex to be described in simple dualistic models. Or to quote Faulkner and Lie: 'One size does not fit all' (2007, p. 172).

Feminist technology research has illustrated how technology and gender are mutually constructed, and we have seen how discursive constructions in recruitment initiatives have different consequences. By looking at the variety within ICT education, a feminist criticism of informatics and feminist technology research, we move the focus away from gender-political initiatives in an attempt to find a new basis for the discursive construction of gender-technology relations.

Variety within ICT education

Questions about what we can learn from a wider spectrum of computer studies in education have only been raised to a degree in the Norwegian debate about 'women and computing', and have barely been mentioned abroad. The trend is that, primarily, computer education within faculties of natural science, engineering and technology, such as informatics and computer science, has received attention due to the problems experienced in recruiting women. Sørensen has pointed out that attempts to include women are often based on studies of the exclusion of women; there are not as many studies of successful integration of women as there are of exclusion (2002b). Knowing the exclusion mechanisms, however, does not secure successful integration; exclusion mechanisms cannot simply be reversed to become mechanisms of inclusion (Faulkner and Lie, 2007).

We have encountered the assumption that women are more interested in the technology's usefulness than they are in purely technical subjects. Already by 1985, Tone Bratteteig and Guri Verne had asked whether women's perspective on informatics could contribute to increased focus on subjects that interested women, and several researchers have subsequently asked similar questions (Bratteteig and Verne, 1985, 1997; cf. Crutzen, 1994; Mörtberg, 1999; Björkman, 2002). However, the main core of informatics or computer science has not changed significantly. Instead, the growing number of computer studies under umbrellas of humanistic and social science faculties have contributed computer education with a greater focus on use, users and the technology's social and cultural consequences, such as we find at the Department of Humanistic Informatics at the Faculty of Humanities, which we will explore further in Chapter 4. By now, it should be clear that more women seem to prefer these kinds of computer-related educational alternatives, as reported in both European countries and the US (Cohoon and Aspray, 2006a).[8] But, we may ask, are the humanistic and social science-based computer studies the 'real thing'? Or is this only another example of how gender dualism has created a discourse of 'computing for girls' and 'computing for boys'? Mörtberg has pointed to a divide that has emerged within computer education in Sweden, between humanistic and social science-based computer studies on the one hand, and computer studies within natural sciences on the other (Mörtberg, 1994, p. 136; see also Mörtberg and Elovaara, 2010). Minna Salminen-Karlsson, who also has studied Swedish computer education, claims that there are reasons to expect that the status of a field decreases as the number of female students increases. At the same time, there is a risk that a field that was earlier defined as a 'technical' field will be redefined as a 'non-technical' field because of the assumed divergence between women and technology (1997, 1999). Thus, a focus on computer education where women actually do participate and seem to thrive is not automatically followed by an appreciation or recognition of women as computer students or computer-literate; we might also find that women entering male-dominated areas have been re-segregated in 'female ghettos', as suggested by Judith Samsom McIlwee and J. Gregg Robinson (1992, referred to by Bagilhole et al., 2008, p. 8).

A focus on the successful integration of women in computer education can probably teach us more about what makes women prefer certain computer-related subjects to others. However, we once again seem to encounter a gendered divide, this time between computing rooted in different scientific traditions. The computer education rooted in natural sciences seems to be the least flexible in relation to gender, but perhaps we can reach further by exploring feminist criticism against the knowledge tradition of computing.

Feminist criticism

It has been emphasized by researchers and computer scientists that computer education needs change, and that improvements could help to 'attract and retain both men and women' (Sagebiel, 2003, cited in Powell et al., 2004, p. 25). But it is also emphasized that change needs to be developed internally (Bratteteig and Verne, 1997; Salminen-Karlsson, 1999; Björkman, 2002). At the same time, the internal perspective has certain limitations, according to Salminen-Karlsson, who has studied attempts to reform educational programmes in technology in Sweden. Salminen-Karlsson found that the technological milieus had difficulties in breaking out of the professional norms into which they had socialized themselves, thus, also having problems in suggesting changes that represented radical challenges to the existing norm within the educational institutions (1999). Also, the professional groups in ICTs departments do not have sufficient knowledge about gender. The result in the Swedish reform group studied by Salminen-Karlsson was that being a woman was treated as a qualification for representing knowledge about gender. Consequently, both the problem and the responsibility for solving it were handed over to women, while men remained both unproblematic and 'ungendered'.

This example illustrates that criticism towards ICTs as a subject needs to be based on knowledge about gender. However, according to Christina Björkman, the distance between informatics and feminist technology criticism has been considerable (2002). The tension between informatics and feminist knowledge traditions has made some female computer scientists distance themselves from feminist criticism, while it has made others 'break with "difficult and impossible" mother disciplines' (Levold, 2001, p. 153). However, several female computer scientists have turned to gender research to search for tools that are useful in challenging the knowledge tradition within informatics and computer science,[9] as we have seen, Bratteteig and Verne advocate in Norway. We remember that this duo already tried to open a discussion about a female perspective on informatics in 1985. Twelve years later they tried to spur this discussion on one more: 'As female information scientists we argue that there are gender perspectives in informatics, and that discussing these is interesting both seen from women research and from informatics. To understand or practice feminist criticism of technology we need to understand what gender aspects are, and how they can exist within technological subjects. There is no discussion about this among our colleagues, not even among our female colleagues' (Bratteteig and Verne, 1997, p. 12). Again, Bratteteig and Verne are trying to use a gender perspective on informatics, pointing in particular to how Sandra Harding suggests that natural sciences should be explored with regard to gender. And, yet again, we can see that the gender perspective has not reached the agenda within informatics. However, this time Bratteteig and Verne seem to find a firmer foundation within the feminist research tradition, and they emphasize the importance of studying 'epistemology, to establish alternative understandings of knowledge' (1997, p.14). They suggest doing this by questioning 'truths' that are taken for granted within informatics, and they start by challenging four topics that they find important for the subject; the computer, information, programming and computer games, asking how these specific topics can be taught in more gender-inclusive ways. Bratteteig and Verne also explore how gender can be brought in as an important issue by focusing on girls and women in the design process. A problem they encounter again and again is that the discussion involves 'assumptions about how women *are*', which easily leads to essentialist conclusions, as we have seen in both the masculine-connoted and the feminized discourses discussed in this chapter. Bratteteig and Verne illustrate the difficulties in challenging a firmly-rooted knowledge tradition, and they conclude that 'sometimes it is perfectly possible to question prevailing perceptions, while it is not quite as simple in other cases (like programming)' (1997, p. 20).

Bratteteig and Verne were primarily focusing on how questions about gender could be involved in the design process by focusing on technology users as both men and women, questions explored also by other feminist technology researchers (Oldenziel, 2001; Oudshoorn et al., 2004). But the knowledge tradition of informatics has also been challenged through a feminist perspective in other ways. Mörtberg challenges one of the central concepts within the main area of informatics: object-orientation. In an object-oriented perspective, computer systems are constructed as models of the world, with the focus on elements that exchange information and services with each other. In this modelling process, a complex world is deconstructed and divided into classes of objects, and, Mörtberg claims, this process also turns subjects into objects. Thus, she challenges this part of the knowledge tradition of informatics by asking whether this could have been described as a subject-oriented design (1999). Mörtberg's suggestion is to make visible the history of the technology, to examine how the involved knowledge has been constructed and to question whose knowledge has been given the opportunity to form the foundation of the professional knowledge tradition. As we saw in Chapter 2, the question of whether women bring something special into the field of computing has also been raised, although without showing clear indications that they do (Sørensen, 1992; Mack, 2001).

'Computing has no nature', Mahoney reminds us. 'It is what it is because people have made it so' (2001, p. 172). And it is not static, although recent changes in content and curriculum are not made with gender questions in mind, according to Bagilhole et al., but rather as an answer to 'industry needs' (2010). However, Bagilhole et al., in their study of curriculum reforms in engineering education in the UK, where hopes of improving the gender balance were tied to increased interdisciplinarity, concluded that 'further curriculum developments in terms of interdisciplinarity per se will have little impact if they are not supported by other reforms to HE [higher education] engineering education culture' (2010, p. 11). This study thus underlines the importance of dealing with the question of women's under-representation in computer education on several levels.

Criticism of the knowledge tradition within computer education is a wide topic, and the aim here is simply to illustrate how this tradition can, and has been, challenged in different ways. One way is to employ a feminist perspective, which will raise the awareness of gender as an important category as well as challenging key concepts and thought structures that are taken for granted within the tradition. And as pointed out by Schiebinger, 'when feminist practices... become widely accepted in science and engineering... they are no longer seen as "feminist", but as "just" or simply "true"' (2008b, p. 9, cf. Chapter 1).

As we have seen, it is a difficult balancing act to focus on gender without (re)producing sharp dualistic or essentialist descriptions of gender. If we want to introduce a greater variation in a discourse of gender and information technology, it is also necessary to question stereotypical perceptions about both gender and ICTs.

Feminist gender research challenging stereotypes

A large and growing tradition of research in gender and technology within the humanities and social sciences has contributed to challenging stereotypical notions of gender–technology relations since the 1980s. Empirical studies of the technology's gendered aspects have contributed knowledge about gender and ICTs that introduces nuances in otherwise essentialist and dualistic gender perceptions. The analysis of computer students we will meet in Chapters 4 and 5 is one attempt to distract the dualistic gender stereotypes. We will see that the students articulate a set of shared perceptions of gender and technology; a hegemonic discourse that seems to dominate and partly suppress other and alternative ways of perceiving gender and ICTs. Simultaneously, they illustrate how real men and women do not always conform to the hegemonic discourse, but rather find their own individual strategies to cope with the discourse; they construct their gendered identity as ICT students in a number of different ways. Thus, they illustrate how the story about the world might be contained within a stereotypical and fixed narrative, in what Laclau would call a 'myth', which is not a precise description of the world, but still a necessary way of retelling reality in a coherent way (Laclau, 1990, cf. Torfing, 1999, p. 115). Such myths still allow for alternative narratives that, to a certain degree, deviate from the myth. The individual narratives of the male and female students, their different positioning strategies and their individual discursive constructions illustrate that the world is not in harmony with a strict dualistic or stereotypical image of gender-technology relations.

Women need female role models within ICT professions, it has been claimed (Björkman et al., 1997; Stuedahl and Braa, 1997; Kekelis et al., 2005; Ashcraft and Blithe, 2010), but how can we create role models when computer competence does not seem to 'stick to' women in the same way that it sticks to men? (Faulkner 2000). We saw Bagilhole et al. pointing out that a change in curriculum alone is not enough, that changes in culture and stereotypes are also required (2010). 'What else can we do beyond putting yet more potted plants in the entrance to the department?', Frances Grundy's male colleague commented to the returning question of how to recruit more women into computing (Grundy, 1997, p. 3). Although Grundy refers to this as a feeling of being 'stuck', Sapna Cheryan et al. would perhaps have claimed that more flowerpots have a mission. They suggest that stereotypes can be changed 'merely through the physical cues present in an associated environment', and that altering the environments can change women's perceptions of 'belonging' in computing (Cheryan et al., 2009, p. 1045). The problem might, however, not only be women's own sense of belonging, but also other's perception of women *as* belonging. 'Women are not considered suitable for jobs traditionally filled by men, not because they are technically unable to perform these jobs, but because women do not carry the symbols, do not correspond to the hero images, do not participate in the rituals or foster the values dominant in the men's culture', Geert Hofestede claims (2003 [1991], p. 16, quoted in Bagilhole et al., 2008, p. 10). And Robinson and McIlwee found that '[m]en are not better engineers, but they are better at *appearing* to be better engineers in a *male-defined* way' (Robinson and McIlwee, 1991, p. 417, quoted in Mack, 2001, p. 160). Femininity and technical skills are not easily associated with each other, and Bagilhole et al. suggest that women who are successful in science, technology and engineering 'are not perceived as feminine' (2008, p. iii). Thus, it is not only a question of how stereotypes have an effect on women's choices of education and career, but also of how stereotypes work on behalf of women once they have chosen a career in technology. Mass media can both reinforce and challenge stereotypes, Jenny Kitzinger et al. point out. However, it is an explicit wish among women working in technology and engineering that, similarly, media provide more diverse representations of women in technological careers, to challenge the stereotypes that disadvantage women (2008).

Boys' and men's fascination with computers has been the focus of a far greater number of studies than female computer fascination (Hacker, 1989; Kleif and Faulkner, 2003; Mellström, 2004). And the expectation of women's low interest in computers in the hegemonic discourse (see Chapter 4) can explain why the feminized discourse relies on attracting women by pointing to the pleasure of working with people instead of working with technology. This, however, misses the important point that women might be equally fascinated and enthusiastic about computing (Abbate, 2010), as we will see in Chapter 5. The feminized discourse is supported by research documenting that girls and women are interested in using the Internet for communication and information (Gansmo, 1998; Nordli, 1998; Håpnes and Rasmussen, 1999; Kvaløy, 1999). Feminist research can, however, also make visible women's interests for other aspects of the technology (Kekelis et al., 2005; Webster, 2006; Alpay et al., 2008). The difference is, as pointed out by Håpnes and Rasmussen, that the Internet, which has been central to the 'communicative turn', was not as strongly associated with men as other aspects of the technology at the end of the 1990s. Positive images of girls' and women's computer use and skills are competing with images of the 'typical computer user' which we saw described as both male and a loner. The same, however, is the case for alternative stories about men. Not all men have extensive computer competence, and these men are also challenged by the masculine norm in the discourse, as we will see in Chapter 4. As with women, they do not fit the specifications for the male norm, and a more in-depth study of this group of discursive 'outsiders' might produce an even more nuanced perception of gender-technology relations.

If the goal is to feed a discourse that accepts a greater variation than the dualistic description of gender and ICTs we have seen dominate the various inclusion initiatives, it is important to document and pay attention to the variation that can be found in both men's and women's relationships to information technology. Another challenge is to produce female role models, including the perhaps biggest challenge, which is not only to make visible the stories about women's positive relationships to and pleasure in technology, but also to make them 'stick' to women, as we will see in the following chapters.

A differentiated discourse

Constructions of gender in the discourse of information technology obviously have an impact on how different subject positions, and consequently different discursive spaces, are made available to men and women within computing. The three perspectives suggested earlier – focus on a broader spectrum of ICT education, internal criticism and feminist technology research – are three ways of challenging the stereotypical discursive constructions of the gender-blind, the masculine-connoted and the feminized discourses of information technology. These perspectives open up opportunities for alternatives and variations in the perception of gender that are ignored and hidden in the three discourses discussed.

In different ways, feminist researchers have articulated theories about gender as a difference that is always in motion; gender as two kinds of a constantly changing 'something' (Haavind, 1994). However, by looking at the different discursive constructions of gender and ICTs, this 'something' is remarkably stable. Women are constructed as either inside or outside the discourse based on the inclusion or exclusion of qualities that are associated with women. The feminized discourse illustrates this most clearly; the novelty of the feminized discourse is neither the idea about women's excellent communicative abilities, nor the assumption about men's interest in 'technical genius solutions'. The novelty is rather that 'communicative abilities' are tied to ICTs. Thus, regarding gender, the feminized discourse can hardly be said to have contributed a revolutionary rewriting of gender and ICTs. Perhaps this is the reason that it was possible to rewrite the discourse of information technology with a feminized twist; it does not challenge the masculine-connoted discourse. Even though men's relationship to computers had to be described in rather negative terms, as useless and unimportant literacy, the connection between men and computer literacy is still intact.

Gansmo has illustrated how Norwegian politicians and headmasters rely on statistical material and their own (stereotypical) assumptions rather than qualitative research in planning gender equality measures in schools (2003). Research producing statistical material is of course important, but unlike qualitative research, quantitative research in this field is most often built upon categories such as girls and boys, men and women, which reduces the opportunity to catch internal differences in the gender categories (Adam et al., 2004). A gender neutral discourse will in most cases mean gender-blind, and will probably not be able to challenge the already inscribed gendered perceptions of computers and computing. A discourse that considers gender could, however, also accept variations, both between men and between women. Hayes, however, suggests that necessary changes in computer science should be created through 'gender-independent approaches... that do not make assumptions about the inherent interests and backgrounds of each gender' (Hayes, 2010b, p. 270). Perhaps Lie's suggestion can be an inspiration, when she suggests we should see 'gender as an "empty shell" which can be filled with desirable attributes and qualities based on personal preference' (Lie, 2003b, p. 277). The main challenge seems to be a general acceptance for a discourse that does not make it a surprise that women like to program or that men have excellent communicative skills – or the other way around. This would in fact be a discourse that is more in line with the empirical studies undertaken since the turn of the millennium of men and women's relationships to information technology, which contribute stories that could support a more differentiated discourse, as in Chapters 4 and 5 we shall see women addicted to the computer, as well as women's fascination for technology. To get closer to the goal, we have to continue to disrupt the dualistic and essentialist perceptions of gender, both on the practical and the discursive levels, and we have to add cultural images that can, at least, exist side by side with traditional and stereotypical images of gender-technology relations (Hayes, 2010b, p. 269).

But, what about the recurring question: 'Why do not women choose computing?' Why should women in Norway, a nation considered to afford its population one of the world's best situations regarding gender-equality, choose to position themselves in traditional and less-acknowledged ways in a structure that also allows them to break out, do new things, increase their wealth, economy, political power and personal opportunities? Charles and Bradley, in their cross-national study of female under-representation in computer science, found 'little evidence that women's representation in computer science programmes is stronger in the most economically developed countries, or that it is stronger in countries where women participate at higher rates in the labour market, higher education, or high-status professional occupations' (2006, p. 194). Charles and Bradley's answer to why this pattern appears is that the most open-minded nations not only provide 'the right to be free from overt discrimination', but also the freedom and 'right to choose poorly paid female-labeled career paths' (2006, p. 196). And from Els Rommes' study of occupational choices among Dutch adolescents, we see that both girls and boys choose highly gender-traditional occupations, leaving technological subjects out of girls' top ten list when considering which occupations they would like to choose for the future (2007). Also, in the US Trauth et al. found that while a majority of boys saw computer science as an attractive college major, this applied to a minority of girls (2010).

Designing strategies to include women in computing not only faces challenges from within educational institutions, but also cultural factors that affect boys' and girls' educational choices. Although women's under-representation in computing might be a global issue, it needs local solutions (Trauth et al., 2009, p. 478), and the local solutions also need to recognize that 'One size does not fit all' (Faulkner and Lie, 2007, p. 172). To produce successful inclusion strategies, it is necessary to recognize variations and differences both between as well as within the gender categories, which we will explore further in the following chapters.

4
Variations in Gender–ICT Relations among Male and Female Computer Students

Introduction

We have seen how perceptions about the relation between gender and ICTs have changed since 1980, and that an impression of stability was created through certain discursive logics that made variation disappear from the dominant stories. We have also seen a parallel movement from a gender-blind to a masculine-connoted and a feminized discourse within computing, creating different positions from which to deal with the 'problem of women'. Also in this chapter we will pursue the question of gender and ICTs in an educational setting, but this time with focus on how real people, young men and women participating in a university-level computer programming course in Norway, perceive the existence of gendered expectations in relation to computers, and also how they deal with these perceptions and expectations. We are also changing focus from the 'traditional' fields of computer education, as the computer course we will study here is one of the 'alternative' courses mentioned in Chapter 3, at the Department of Humanistic Informatics. In Scandinavia, as reported elsewhere, there is a tendency for computer-related education within social sciences and the humanities to attract more women than corresponding courses under the faculties of natural sciences, technology or engineering (Corneliussen, 2003b; Ogan et al., 2006). As the name indicates, Humanistic Informatics is a computer course within the humanities, at the Faculty of Humanities. More than 60 per cent of students were female in the late 1990s,[1] a number that also reflected the proportion of women at the Faculty of Humanities, supporting Cohoon's finding that more women are recruited to computer courses 'when their home institution enrols more women' (2006, p. 229).

The subject offered at this department can be described as the 'soft' side of informatics, as critical perspectives on information technology are important, involving topics of social, cultural, ethical and aesthetic aspects of information technology, with focus on uses and users of technology. But it also includes 'hard' topics, such as technology development and computer programming, which at the time was considered one of the major fields in the study programme.[2] Students graduating from this department have reported that they are working with technology in different ways; as webmasters, as web developers or designers, programmers, computer games developers, e-learning specialists, media lecturers and net communicators. Thus, many of the students from Humanistic Informatics have found work in fields referred to as the 'new' ICT professions (Valenduc et al., 2004, p. 10), or ITEC, Information Technology, Electronics and Communications (Webster, 2006, p. 7). They might also serve to illustrate Faulkner and Lie's claim, that 'the boundary between using and creating ICTs is now very blurred' (2007, p. 173).

In this chapter we will follow students at the Department of Humanistic Informatics through their first three months at the initial computer programming course. Three phases in the cultural appropriation of computers were recognized in Chapter 2, and the study this chapter reports originates from the height of the second phase – the period when the masculine discourse was most solid. Thus, this chapter will explore how men and women deal with a hegemonic discourse that very clearly creates not only different, but also more favourable expectations of men than of women in relation to computers, which as we have seen in the previous chapters is an issue haunting more or less the entire Western world (Cohoon and Aspray, 2006b; Phipps, 2007). Contrary to the homogenization of perceptions of gender-technology relations we saw in Chapter 2, we will see far more variations among both men and women, and the chapter will illustrate opportunities and limitations in discourses on gender and technology.

Despite the image of Norway as a country of gender equality (cf. Skjeie and Teigen, 2003), very few women choose to study or work with information technology. One of the problems reported by women within computer education has been related to being a minority in male-dominated institutions (Håpnes, 1992; Stuedahl, 1999). Efforts to attract more women to computer education have shown temporary positive results, but have not yet proved to have long-lasting effects (see

Chapter 3). Although it is true that when women are a minority, their experiences of the social setting may be negative, the persistent low number of women completing degrees in computer science indicates that the problem not only has to do with male dominance in numbers, but also indicates the importance of investigating other barriers women might experience in their relations with computers. To find out more about why the number of women fails to increase, in particular in subjects related to computers, we need to learn more about how men and women experience themselves in relation to computers. We need to ask what it means to be a woman working with computers compared to being a man working with computers. In short, we need to explore how gender and technology interact and affect each other in lived lives.

Studies focusing on female computer students tend to find them in a social context dominated by male students and employees. One of the questions raised by studies of women in male-dominated environments is whether they were 'ordinary girls' (Edelman, 1997), or represented a special group of women who had made extraordinary educational or career choices. With more than 60 per cent students being female, a percentage reflecting the general gender distribution at the Faculty of Humanities, the students at the Department of Humanistic Informatics can hardly be seen as more special than the average student at this faculty. Thus, it is possible to assume that the female students in this study do not feel they belong to a marginalized group, as was found in other Norwegian universities in male-dominated computer departments (Håpnes, 1992; Stuedahl, 1999). The female majority at Humanistic Informatics makes this a different environment from the male-dominated computing subjects, often reported to be – at least symbolically – dominated by male computer nerds or geeks. The young men with a more than average interest in computers have been found, for instance by Margolis and Fisher, to be one of the reasons some female students reconsider their choice of study and end up seeing themselves as 'just not interested' (2002, p. 78). Thus, the majority of women in Humanistic Informatics makes this an interesting context in which to study whether gender makes a difference to individuals' engagement with computers in other ways than by men outnumbering women.

To understand the 'problem' of women in relation to technology, we also have to understand men's relationships to technology, Faulkner claims (2006, p. 1). Grounded in a similar belief, both genders are included in this study, where I followed seven men and 21 women participating in a programming course for first-term students that I was also teaching (Corneliussen, 2002).[3] Their previous experience with computers varied, and while some had their first computer as a child or in early youth, others had been introduced to the computer more recently. Most of them had used the computer for a number of different things, and school-related activities clearly dominated among both men and women, followed by email, Internet and computer games. There were no noticeable differences between men and women in terms of experience and activities, but considerable differences within both groups. The same goes for their confidence in working with the computer, as could be observed in the computer lab. All of them became more advanced computer users during the course, and with that, as we will see, perceptions of computers as well as motivation changed for many of the participating students.[4]

During the 10 weeks of the lectures and computer lab exercises, I sent questions via email to the participants asking about their experience with computers, both in and outside the educational context, and both with and without matters of gender included in the questions. At the end of the term, most of them were interviewed in groups of three or four.[5] In addition to emails and interviews, throughout the term the computer lab was an arena for observing the students working with computers as well as with each other. The time spent in computer labs with the students provided a good opportunity to get to know all the students individually, and the students also responded very positively in these classes, remarking on the comfortable atmosphere. One of the students explained how there were 'lots of shouts of joy' and pleasure when a code worked, or even laughter when a code gave unexpected and funny results. This triangulation or multi-method approach made it possible to get to know the students very well, to get a wide perspective on their thoughts about the topics I raised, as well as an insight into how these topics were discussed among the students.

The theoretical perspectives of the following analysis is in line with the theoretical frame discussed in Chapter 1; both gender and technology are seen as categories that are not stable or fixed, but rather are constantly (re)constructed in interaction with each other. Scott suggests that gender should be investigated as a discursive category based on 'perceived differences between the sexes' (1988, p. 42). De Beauvoir adds the important notion that we all contribute to the construction of gender, through what we do about what the world does to us (de Beauvoir, 1989 [1949], referred in Moi, 1999, p. 72). It is this dual understanding, of what the world does to us and how we deal with it, that we will investigate here. Discourse theory (see Chapter 1) is important also to the analysis in this chapter. However, here we will not only focus on 'discourse', but also on 'subject position' (Laclau and Mouffe, 1985, p. 115), representing a discursive point of identification within a discourse. Thus, we are going to explore how individuals use the discursive structures and the accompanying subject positions as a frame of reference for their own articulations. While considering discourses as temporarily fixed truths which are both influencing and influenced by individuals' articulations, it is possible to see the articulating individual as both shaped by and a producer of the social reality around us. Thus, we will see how the individual can associate with or negotiate a subject position. This perspective gives the opportunity to see the individual as an agent in society, or what Søndergaard calls an 'active subject' (2000, p. 67).

Because gender is something we can 'perform' without conscious reflection, it can be difficult to study people's experiences of being a man or a woman, and there is considerable risk of obtaining answers such as 'I haven't thought about that', or 'I don't think that gender makes any difference'. I received a few answers like that, but the informants were in general very helpful in wanting and trying to provide answers to questions about gender. It is also difficult to discuss gender in a cultural context in which gender equality through legislation is set as a standard for society. Even though it is possible to find examples of gender inequality in Norway, the general goal of gender equality makes it more 'political correct' to claim that gender does not mean anything; gender is not supposed to result in men and women being treated differently. This has even produced a common 'myth' about gender equality in Norway that seems to hide continuous gender inequality (Skjeie and Teigen, 2003), which is an effect of gender mainstreaming politics also found in other Western countries (Godfroy-Genin, 2009, p. 85). Thus, when the computer students we are going to meet here claim that 'gender does not mean anything', they do not necessarily refer to how things are, but rather to how things should have been.

A hegemonic discourse of computing

If you enter a store to buy a computer, most of those working there are men. In most of the large computer companies, it is mostly men you hear about, or see speaking about things. So the main impression I have is that the computer culture is run by men. (Female student)

Girls, they sort of haven't got a clue – at least many of them – about computers. (Male student)

They [boys] sort of don't need to do anything like this [a computer course] to be just as good as us. (Female student)

These quotations are from three of the computer students, and they illustrate not only how computer technology is associated with men, but also that this is something we 'just know'. We know that men dominate many contexts where the computer is central, but except for the male dominance in numbers, what is it that makes these students assume that boys and men have an invisible connection to computer technology that makes it easier for them to learn to use the technology? And how do they know that girls 'haven't got a clue'? This 'knowledge' is part of the context to which both male and female students are invited when they enter computer courses. How does the gender difference implied in this 'knowledge' make a difference between being a man or a woman within computer education? How can women find a place of their own in a field so strongly associated not only with men as the large and dominant majority, but also with men's abilities? And is it really true that it is easier for men to establish a relationship with computers?

As we saw in Chapter 2, the masculine discourse that was consolidated throughout the 1990s includes gender as a primary difference in relation to information technology. Many of the ideas from this masculine discourse are also shared by the male students in Humanistic Informatics. One of the most important expectations, both in itself and because it produces a range of other gender differences, is about men's and women's different levels of interest in the computer (cf. Nissen, 1996; Gansmo, 2003; Håpnes and Rasmussen, 2003; Lagesen, 2003). While boys and men are assumed to have an intense interest in the computer, girls and women are not expected to share this interest. Men's greater interest is also affecting expectations about knowledge and skills related to computers. Women's (assumed) lack of 'technical insight' is suggested as one important barrier for women, and one of the male students in the department associates girls with 'not having a clue' about computers.

The expectation about women's lack of interest does not, however, mean that women are assumed not to use the computer. The computer is a multilayered technology used for a number of different activities, and men and women are assumed to use the computer for different purposes. Boys and men are associated with computer games, and with technical skills necessary to make the computer work. Women, on the other hand, are expected to be more familiar with simple 'use of programs', and using the computer for a reason, different from men's inclination to explore the technology for itself (Turkle, 1984; Lockheed, 1985; AAUW, 2000; Kleif and Faulkner, 2003). Thus, one of the main dividing lines seems to go between perceptions of men as the technicians who are able to understand and fix the machine, and women who only use the computer for a purpose and have less depth of knowledge about it. The computer is, however, 'something more', one of the male students claims, and there are some fields that the male students associate with women, such as communication, design and aesthetics.

The female students share most of the male students' perceptions, although they seem to be more aware of the male dominance in numbers, contributing to a masculine gendering of the computer culture. They wonder whether 'men can acquire knowledge easier...because they have more interest' in the technology. In accord with the male students' opinion, women suggest that men can learn more quickly. However, they question whether this is really true, or rather 'something that boys want girls to believe?', introducing the possibility of a cultural deception or delusion, making girls and women believe that men are the masters of this technology.

The female students assume that boys have a more playful and pleasurable relationship to the computer than girls, making them more reckless, 'daring to explore possibilities with computers'. This resembles a 'risk taking as a learning strategy' which Sherry Turkle found among boys, claiming that this gave them an advantage in acquiring computer skills because they were not afraid to break things (1988, pp. 47–9). The male interest in computers is also associated with friendship and fellowship among men – a fellowship of which women are not part, as was also found by Kleif and Faulkner (2003). Several of the women also find men's constant talk about computers unbalanced and unhealthy, and different from their own more balanced way of approaching computers.

The discursive space for men is described as a far more advantageous space than the discursive space for women. One of the effects of the masculine discourse is that the presence of women is not anticipated in computer-related contexts. Even women who had themselves chosen a computer course were 'a little surprised that there were mostly girls on the course'. One of them even suggested that 'people are surprised when a girl actually find computing fun and chooses a subject like that!' The presence of women as lecturers also surprised the women. However, the effect it had on an otherwise masculine discourse varied. Some of the students claimed that it did not affect their perceptions at all because I (a woman and lecturer in programming) had 'entered a male-dominated world'; while another female student claimed that I had 'neutralized' it a bit, although, only in the local context of Humanistic Informatics, and not in general.

The majority of women among the students is used as evidence that women both can and want to learn about computers. Some of the women also expressed relief in finding so many women on the course; it made the choice of a computer education less 'weird', illustrating the fact that numbers matter (Faulkner and Lie, 2007; Lagesen, 2007) and the importance of a 'critical mass' (Kanter, 1993 [1977]) to avoid the minority gender perceiving the field as 'gender in-authentic' (Kleif and Faulkner, 2003, p. 310). With between 60 and 70 per cent of the students being female, Humanistic Informatics obviously had a 'critical mass' of women. This was, however, not equally comforting for all the women, as women in general are still in a minority within what they referred to as the 'real' computing studies. Even though the female dominance on this course is an obvious evidence that 'women want to and women can – also in IT', the female students interpret this as if women only want 'a little' computer knowledge, and, compared to male students, as if women only *have* a little knowledge.

Despite variations, the hegemonic discourse is used as a valid description of gender and computing in general, and this discourse has two main subject positions; one associated with men and one associated with women. The two subject positions of the hegemonic discourse appear as opposites; features characterizing one do not characterize the other, and presence in one becomes absence in the other. The subject position associated with men includes interest in the computer, and playing with the computer can be a goal in itself. The male position is associated with both leisure activities and play, such as computer games, and with exploring the computer and programs on a more advanced level, similar to assumptions about men's relationship to technology in other Western countries (Nissen, 1996; AAUW, 2000; Mellström, 2004). Thus, the subject position associated with men implies an expectation about interest in, experience with and knowledge of computers. However, the subject position associated with women includes an expectation, first, that women's interest in computers is limited, and second, that in order for women to develop an interest in computers they need to find them necessary or useful for certain purposes (Faulkner and Lie, 2007). The computer is primarily assumed to be a tool for different objectives for women, rather than the target of interest itself. The 'technical' or physical computer that is important in the male subject position is not only assumed to be uninteresting in women's relationship to the computer, but also more or less excluded from the subject position associated with women, and women are certainly not expected to undertake computer repairs that require tools to open them up and deal with their physical parts. The position associated with women thus creates expectations that women's experience with and knowledge of the computer is limited, and that the wish to use a computer is tied to certain objectives (Aune, 1996). While men are associated with an aimless, yet, invaluable exploration of the machine, women are associated with a rather limited 'user relation' and with applications for text, Internet, communication and information (Håpnes and Rasmussen, 2003). Similar perceptions of women's relationship with computers have been found in other Western countries, and have caused anxiety about women's position in 'contemporary e-culture' (AAUW, 2000, p. x). The perception of women's relationship to computers is most notably a negation of the dominating image of male relations to computers; expectations of women's relationship to the computer are best described as a lack or deficiency compared to the subject position associated with men. The only fields of computing in which women are assumed to have advantages, such as communication, design and aesthetics, can all be described by nontechnical words, while boys and men's activities and preferences are all associated with knowledge of the computer itself.

This description of the students' perceptions of the gender–ICTs relationship should not be understood as a description of real men and women, but as a description of how the discourse is creating expectations of men's and women's relationships to computers. This has similarities to how boys' and men's relationships to the computer has often been described as playful and explorative (Turkle, 1984; Levy, 1984; Nissen, 1996; Mellström, 2004), parallel with descriptions of girls and women's relationships to computers emphasizing the usefulness of technology, and of seeing the computer as a tool rather than an end in itself (AAUW, 2000; Håpnes and Rasmussen, 2003; Faulkner and Lie, 2007). We have seen repeatedly in this and the previous chapters how masculinity and technology are linked in cultural stories, similar stories to those Bagilhole et al. found to be accepted in society at large in the UK (2008, p. 8). This is an important foundation in the hegemonic discourse found in this study, and it is the subject positions defined in this discourse that men and women in this study relate to when they construct their own positions as computer students. However, they negotiate it in very different ways, as we shall see in the next section.

Positioning strategies

The articulations from the students include stereotypical and dualistic descriptions of gender and computing, and these support the hegemonic masculine discourse and the expectations it creates of men's and women's relationships with the computer. But do the male students meet the expectations described in relation to the subject position associated with men? Are women really uninterested in computers? The effects of the discourse need to be studied through 'lived life'; through the students' own experiences of being a man or a woman on a computer course. In the following we will explore these students' relationships to computers and the way in which male students and female students handle what is described as a hegemonic discourse. Or to follow Simone de Beauvoir, what they do with what the world does to them.

The hegemonic discourse has a certain power in that all the students refer to it. But the individual's freedom to negotiate a discourse is also illustrated in their descriptions of their own relations to the computer. They introduce nuances and details, and make their own meaningful connections in order to describe themselves as 'understandable' in relation to computers. Thus, we will see how they exemplify a number of possibilities through what I will refer to as positioning strategies when they construct their own relations to computing and computers.

All the students use the hegemonic discourse as a starting point, framing their individual positioning strategies. They are, however, aiming in different directions, according to whether they display harmony, disharmony or rejection of the hegemonic discourse. Thus, by focusing on how the students position themselves in relation to the hegemonic discourse, it is possible to see a pattern of seven different positioning strategies, three among the men and four among the women. The presentation that follows starts with the two most harmonious strategies, and continues with two strategies that are on the move in relation to the discourse. This is followed by two strategies that more or less reject the discourse, and finally, one strategy based on acceptance of the discourse, however, doing so through an association with the opposite gendered subject position.[6]

Men and women in harmony with the hegemonic discourse

The first two positioning strategies are notable for the way they relate to and confirm the hegemonic discourse's masculine and feminine subject positions. The students articulating this positioning strategy use gender to activate a set of gendered associations that explain their own relationship to computers. The discourse creates very different starting points for men and women. Thus, even though they articulate harmony with the hegemonic discourse, the results – the gender-technology relations they associate themselves with – are very different. Two men and two women in the group of students represent these positions, the first illustrating men who are 'rooted in a room for men', and the second 'women in a limited room' for women.

Rooted in a room for men

The two men we will meet first appeared to be firmly rooted in the hegemonic discourse's room for men; they had considerable experience with computers, and interest and knowledge that allowed them to portray themselves in harmony with the masculine position of the hegemonic discourse. They referred to how their long-term and positive relationship with computers was a product of being a boy or man: 'as a boy I have been involved in computing of some form or other since I was in elementary school'. The link between boys and computing is so close, one of them claims, that people 'can almost see' that he is a computer student, illustrating how gender activates a certain expectation about his computer interest. Another man finds the computer course particularly useful because it formalizes the knowledge he already possesses, thus pointing out how he 'naturally' belongs there. Their previous experience with computers is helpful in the study they are attending, not only directly because of the knowledge they already possessed, but also indirectly, as an assumption about 'technical interest' usually being greater among men. Even though they associate themselves with an image of computer-literate men, both of them distance themselves from the one-sidedness associated with computer nerds, or men with a one-sided, purely technical interest in the computer, rather than the useful aspects. That is, however, an old stereotype, one of them claims, and today, 'the computer is out of the boy's room' – it does not require the same effort to make the machine work any more.

They display a certain distance from the masculine subject position, and simultaneously, it is exactly this male image they refer to when pointing out their advantages as computer students. Consequently they seem to benefit both from the masculine stereotype of the technically-interested men and from perceiving themselves as more 'modern' by rejecting this one-sidedness. Thus, while being associated with certain aspects of the masculine position is considered negative, the position itself provides a set of easy-to-use associations between men and computer knowledge. This brings to mind how Robinson and McIlwee found men not to be better engineers, but to be 'better at *appearing* to be better engineers in a *male-defined* way' (1991, p. 417, quoted in Mack, 2001, p. 160).

The men articulating this strategy provide descriptions of themselves that correlate with the hegemonic discourse; they emphasize early and extensive interest in computers, a special male technical interest, and male bonding around computer games. Their distance from the most extreme images of one-sided male fascination with computers does not prevent their association with the masculine subject position. They are rooted in the 'room' that the discourse provides for men, and they use the discursive expectations of men as positive descriptions of themselves.

Women in a limited room

The hegemonic discourse's subject position associated with women was described as *limited* or *deficient* compared to the masculine subject position, and the women we are about to meet aim for this 'limited room' for women. One of these women, a little older than the average student, used a computer for the first time when she was 11, and she had been employed for 10 years in a job involving computer use. Despite her long-term experience she 'was not very enthusiastic about the machine'; she described herself as 'a bit afraid' at the start of the term. After three months, this had changed, and she could even 'miss the machine' when being away from it for more than a few days. However, she found the computing course very difficult, and she was both surprised and disappointed that her previous experience had not been more of an advantage to her. She thought she had some kind of short-coming in relation to computer knowledge, and when explaining why she found computer programming hard she referred to her gender: 'Because I'm a woman and don't understand any of those [computer programming] things'. It was as a woman that she failed to understand programming, and she used 'woman' as the keyword to activate the association of a negative relation between women and computer literacy. In this way, gender itself becomes the explanation that makes it natural that she does not understand computing. The other woman describing herself as having a limited relationship with computers made her own chain of equivalence to illustrate her distance from computers: 'computers = technical = boy things = boring', she wrote to me in an email. Similar perceptions among girls have been found also in other countries, for instance by Kekelis et al., who in interviews with girls in the US found that '[m]any girls believe that it might be boring to work with computers' (2005,

p. 100). However, they also found evidence that influence from family members working with technology could make girls refer to exactly the same arguments, 'like "working hard" and "sitting at the computer"', as reasons why computing would be interesting to work with (Kekelis et al., 2005, p. 106). These various findings thus indicate that perceptions of computers are closely connected to available cultural images.

The women in this group, however, appear to have a rather traditional perception of gender-technology relations, and the dividing line between men and women in relation to computers seems to characterize most of their articulations. They find it 'obvious that boys have a better understanding for the technical stuff', and they are disappointed by how lecturers do not adjust their teaching to women who take longer to understand, relating that also to gender: 'It is after all men who teach ... or women with very high education!' Earlier research has shown that girls and women can feel alienated or excluded because of boys' and men's use of computer terminology (Vestby, 1998; Stuedahl, 1999), but it is not terminology that causes the problem for these women. Rather, it is a problem of being presented with too much or the 'wrong kind' of information. 'I have learnt a great deal ... perhaps a bit too much', one of them claims. Although they were both eager to learn, they had definite ideas about what they wanted to learn. They kept questioning why they had to learn the various topics – 'Where the heck are we supposed to use that?' – and they found many of the technical subjects highly irrelevant: 'I am not going to be an engineer. I am not going to poke about in the machine at all – I don't get anything of that. And I know what it looks like inside ... but those wires and all that: no, forget it. I don't care about that. I don't understand anything ... I don't want to understand it either. I don't want to learn it!' Powell et al. found similar reactions in their study of higher education in engineering in the UK, with women questioning the importance and usefulness of parts of the study content (2004, p. 28). Powell et al. suggest that 'the culture and structure of the engineering education system has [sic] been designed for a male audience' (2004, p. 21). However, as we will see in the next section, not all women reacted negatively to the content of the study in Humanistic Informatics, and some of them rather appreciated the same things that these women rejected as 'too much' and irrelevant.

Both women in this group, which aims for a 'limited room for women', rely on gender when describing their own relation to the computer. However, both of them also at some point reject that gender is the explanation for why they find computing difficult. This simultaneous acceptance and rejection of gender as the differentiating feature should, however, not be read as an inconsistency, but rather as refer-ring to two different discourses. The discourse of computing explains why they, as women, find the study hard and partly uninteresting, while the discourse of gender equality to a large degree has made it politically incorrect to talk about gender as resulting in differences in individuals' abilities, in particular academic abilities. Thus, both women illustrate a positioning strategy aimed at describing them in accordance with the female subject position of the discourse of com-puting. They are women encountering a masculine field, and they use the discursive description of women's limited relation to computers as a valid description of themselves. Or rather, referring to gender *alone* is enough to activate this particular image of women and computers. In short, being a woman is what makes it comprehensible that they have limited interest in and knowledge of computers – in rigid conform-ity with the hegemonic discourse. Again, we can sense the echo of Robinson and McIlwee's claim that men are better at appearing to be good with technology (1991), or Hofstede's claim that women 'do not carry the symbols' required to be perceived as part of 'men's culture' (Hofstede, 2003 [1991], quoted in Bagilhole et al., 2008, p. 12). Several studies have illustrated that women are not necessarily perceived as suitable, 'proper' or skilful students in technical subjects. Carter and Kirkup even found that male students did not take female students seriously, but rather assumed they were only looking for a husband (1990, cited in Bagilhole et al., 2008, p. 21).

It is worth noticing that, despite the majority of female students or female lecturers in Humanistic Informatics, this did not have any partic-ular effect on these two women's perceptions of computing as a mascu-line field. The male students, despite their minority position, dominate in their experience of the classes, and the female lecturer (me) did not become a positive role model, but rather turned into a trans-gendered figure, absorbed by the masculine culture. Thus, it seems that their ini-tial perceptions of computing as masculine constrained the opportu-nity of a non-traditional situation with a majority of women having an impact on their perceptions.

On the move

The next set of positioning strategies illustrates men and women who are about to change in relation to the hegemonic discourse. A group of non-hegemonic men work to achieve the masculine subject position – they aim at a room for men, while a group of hegemonic women strive to expand the limited room for women, aiming at a more open room for women. Thus, these men aim at the hegemonic discourse, while the women depart from the hegemonic discourse.

Aiming at a room for men

The two men exemplifying this positioning strategy could not exhibit the same harmony with the masculine subject position as the first group of men we met. Even though one of them had already used a computer as a child, none of them had continuous experience with computers. The computer course, however, appeared to be a positive experience, finally giving them greater knowledge, and one of them described it as 'quite cool' to finally learn more than 'ordinary people'. Despite their lack of previous knowledge, they both suggested that they have an advantage compared to women. One of them thought that women 'appear to be "stupid" or "slower" than men' in a male-dominated setting because men and women communicate differently – ignoring the fact that the course he attends is dominated by female students. He was, however, aware of this and it both frustrated and comforted him. First, the women frustrated him because they kept asking for his help with the computer, but did not follow his instructions because they wanted to know 'WHY', he explained to me in an email. Thus, he seemed to have been associated with the hegemonic masculine subject position, even though this position involved expectations of knowledge that he did not have, which made the women's requests for explanations problematic. He illustrated how he had access to the masculine subject position, even though he lacked the background and knowledge required to fulfil the expectations of this position. He was, however, not unaware of this ambivalence, as he found comfort in this environment with a majority of women: 'I think it is good that there are so many women in Humanistic Informatics. Somehow it makes me feel more confident. ... If men had been in the majority, you would have to (I feel) pretend all the time that you knew more than you actually do, in order not to appear "stupid".' He seems to express relief about not having to be tested as qualifying for the masculine position by virtue of other men's presence. Even though he is aware that he does not qualify for the masculine subject position, he stays in it as long as it remains unproblematic, illustrating the readily available associative link between men and computer knowledge, as with one of the men we met earlier. For this man, however, his gender identity appears to be a double-edged sword: it makes it possible for him to be perceived as computer-literate while simultaneously putting him under constant threat of revealing his less than perfect abilities.

The other man illustrating this strategy also lacked the required 'boy's room' experience, but he was very eager to learn and was one of the students most actively seeking to expand his knowledge and computer skills. In a joking tone he claimed that he wanted more knowledge, aspiring to be 'like a hacker'. Despite his lack of experience with computers, also he assumed he had an advantage because it would be ' "taken for granted" that computers-are-something-I-can-handle, because I am a boy'. Thus, he highlights how expectations of boys' computer interest and competence itself become a positive motivation for his own learning (cf. Turkle, 1984).

Both of these men are on the move in relation to the masculine subject position. They both lack the 'proper' experience and skills, but as we have seen, one of them illustrates his access to the masculine position despite this lack, and the other strives hard to achieve this position – or an even more extreme variant associated with hacking. They illustrate that men, as 'male marked individuals' (Staunæs and Søndergaard 2008, p. 152), are expected to possess computer knowledge, as well as how they as men can readily be associated with computer knowledge because of these expectations. The effect of these expectations can, however, be either negative (through the constant threat of being exposed) or positive (through the cultural expectation that computing is something they can easily learn).

Creating a more open room for women

The largest group of women (a total of six) exemplifies a strategy in which the women adhere to the hegemonic discourse as simultaneously, primarily based on their newly acquired experiences and skills, they redefine the subject position associated with women. They aim at creating a more spacious room for women, though still within the frame of the hegemonic discourse. However, it is not primarily limits in women's relations to computers that characterize this positioning strategy, but rather an emphasis on women as different from men. The distinctive feature making this different from the strategy aimed at a 'limited room for women' is their insistence that women do engage in computing, but they persistently do it 'as women', in ways that are limited compared to men's engagement.

All the women in this group had some previous experience, but all of them also described themselves as unskilled, or 'illiterates'. During the course they all started to feel more confident and pleased about everything they had learnt, in particular about the 'engineering stuff' that the first group of women described as uninteresting and 'too much'.

These women, however, wanted to learn 'everything', and many of them described a newfound pleasure, and even addiction, in relation to computers (see Chapter 5). They were positive about the course in general, describing it as 'practical, not only theoretical', 'very informative', and 'finally a subject we can use'. This group of women found positive support in the female-dominated milieu, acknowledging female lecturers as positive role models, unlike the first women we met. The high proportion of female students became proof that 'women can', and made it less strange to be a female computer student. Thus, they were eager to learn and they did learn, and to their surprise, they started enjoying it!

They did, however, draw a sharp line between themselves as women compared to boys and men. Three of these women were interviewed together in a group, and two of my questions made them burst out laughing: first, the question of whether they fixed their own computer – which they rejected, and second, when I asked what they preferred to do when they were having fun with the computer. The last question produced a roar of laughter, making them repeat 'having fun' in a mocking voice, as if I had said something highly unlikely, while explaining to me that they did no such thing. Boys, on the other hand, might have fun with computers, they explained. Women would not sit aimlessly with the computer, but needed to have a goal or a task, they claimed, and described themselves as 'typical girls', unlike the boys who can 'just sit there', open the machine and have fun. They held onto the difference between girls and boys in relation to computers. In their identity constructions as female computer students, they emphasized that they were *not* doing the same things that boys do, and these negations became vital in being a 'typical girl'. The fact that they played computer games and explored things on the machine themselves did not change their perception of this as a 'boy thing', as they also drew lines for how *much* they were engaged in activities associated with boys. Thus, despite their new enthusiasm for computers, they fell in line with the claims made, for instance by Turkle and by Håpnes and Rasmussen, that being a girl involves distancing themselves from the image of men's intense relationship with computers (Turkle, 1988; Håpnes and Rasmussen, 2003). However, as Webster points out, simultaneously as women distance themselves from 'men's "love stories"' with computers, they 'can be strongly attached to technological work' (2005, p. 5, cf. Faulkner, 2005). Thus, drawing a line between men and women is not automatically followed by a lack of interest.

The women in this group all perceived computing as a male field, in the same way as has been found in other Western countries (Mahoney,

2001; Adam et al., 2005; Faulkner, 2006; Bagilhole et al., 2008). They assumed men's and women's relationships with the computer were different from each other – in keeping with the hegemonic discourse. They saw themselves as 'typical girls' in relation to computing as a masculine field. Compared to other women, however, they call themselves 'non-traditional women', breaking traditional gendered borders. A similar scepticism regarding other women among women who have themselves entered a male-dominated field of technology has been found in other studies (Nordli, 2003; Godfroy-Genin and Pinault, 2006), again illustrating how cultural images are not easily modified, not even by real-life examples (in this case themselves) that are not in harmony with the hegemonic discourse.

One of the women in this group described her previous experience in light of being the 'stupid big sister', too afraid to make a committed effort with the computer. And her lack of computer skills is also what motivated her to apply for this course: 'I did not want to keep being the illiterate I felt I was'. Being an illiterate is not only to be unable to read and write, but also carries a symbolic meaning of not being able to participate in modern society. Thus, she illustrated how she started out as an outsider to the computer culture, but simultaneously she was willing to claim a position within it, as were the other women in this group. And the pleasure they expressed largely referred to how they acquired knowledge in a *male* field. They expressed this as having entered 'a world that's a little bit forbidden', and 'feeling a bit masculine'. Different from Mack's claim that 'young women often do not want to be seen as masculine' (2001, p. 161), or Godfroy-Genin's German informant claiming that engineering 'is not fit for a woman' (2006, p. 31), these women reported in an assertive voice that some of the authority of the masculine field of computing was rubbing off on them. Thus, they appeared to experience their new computer skills more in keeping with Bagilhole et al.'s findings, of women experiencing it as pleasurable and empowering when 'making it in a "man's world"' (2008, p. 33).

The women in this group illustrate how they started near the hegemonic subject position associated with women; with a limited relationship to computers, as outsiders and illiterates – if not in terms of practical knowledge, then certainly in terms of perceptions of themselves. They are, however, breaking some barriers by attending the computer course: they are on the move; they want to learn and they have started to expand their space. Thus, it is as 'typical women' that they engage in computing as a masculine field, but compared to other women they are 'non-traditional' (cf. Adam et al., 2005; Powell et al.,

2006). They have started to cross a gendered border of which they are well aware: they have entered a (forbidden) masculine field from which they gain status and prestige. They have stopped being illiterates and are about to acquire knowledge that will enable them to participate in conversations 'with men who use computer terminology'. They cross the gendered boundary, but they do not break it down. It is still a masculine field, and they still do not perceive themselves as doing as much or having as much fun as boys and men.

Outside the discourse – distance and rejection

The next two positioning strategies illustrate how both men and women can distance themselves from the hegemonic discourse. However, once more the different starting points affect the outcome, as in some ways they appear diametrically opposed to the two strategies in harmony with the hegemonic discourse in which men associated themselves positively with computer skills, while women did not. The strategies we will meet are, to the contrary, about placing distance between themselves and the hegemonic subject positions – a man rejecting having a close bond with computers, and women insisting on being recognized as computer-literate.

Rejecting the masculine subject position

The third positioning strategy found among the male students was articulated by only one man; a man who does not want to be associated with the masculine subject position. Instead, he repeatedly emphasized that he lacked experience and knowledge of computers. He was aged over 20 the first time he used a computer, thus he did not have the boy's room competence associated with the hegemonic masculine subject position. In the computer lab he seemed insecure and a little reserved when working on the computer. My impression of him as unfamiliar with computers was further strengthened as he several times spoke of himself as poorly skilled, using his non-working PC as proof of his lack of technical skills. Thus, he limited the description of himself as computer literate, and when talking with other people he tended to avoid telling them that he was a computer student: 'I cannot be bothered to talk about it'. Thus, he both positioned himself and appeared as a man with little computer experience and knowledge. At the end of the term, however, despite having been questioned about his previous computer experience at an earlier stage of the course, he surprisingly said that he had in fact studied computing at another department at the university before. Thus, he not only seemed to emphasize his lack of skills, but omitted stories about his previous computer experience.

This man clearly did not want to be associated with the masculine subject position, and he illustrated how he needed a specific strategy to avoid it. He was definitely not among the most skilled men, and perhaps his strategy intended to avoid general expectations of men's computer skills, as we have seen how this association (too) easily is available to men based simply on gender. Thus, also, this man apparently had access to the subject position of the hegemonic discourse, but instead of using his qualifications, for instance his previous computer courses (or gender alone, as some of the other men had), his articulations showed that he tries to disqualify himself for the masculine subject position. The hegemonic discourse, however, continues to be a story of computer-literate men and men that are interested in computers, as he adjusted the expectations of himself and claimed that the hegemonic discourse was not valid for all men; it was not valid for him.

A common room for men and women

The third positioning strategy among women has similarities with the previous strategy, as this also involves a distance from the hegemonic discourse. However, while the man we met in the previous section tried to distance himself from competence in computing, the women pursuing this strategy were aiming at a position within computing, in what they see as a room in which gender should not define who should be included or excluded. The women exemplifying this strategy were in general among the youngest students, and they had experiences in line with the 'boy's room' experience, with computers and computer games being a natural part of their environment since they were children. Unlike the woman who saw herself as the 'stupid big sister', one of these women expressed her capability by describing how she was the one that fixed her brother's computer.

Thus, the women in this group had broad experience with computers, and in general they rejected the idea that gender affects what kind of relationship people have with computers. That does not, however, mean that they perceived men and women as similar – they saw clear differences in how men and women use the computer. One of them explained how men have an unhealthily intense relationship with computers, while she, even as a computer student, did not 'have to talk about computers all the time'. The new space she was aiming at has a clear gendered border, but simultaneously it has room for a more positive subject position for women.

These women rejected the hegemonic discourse primarily by rejecting the limited room for women. That does not mean that the hegemonic discourse does not exist – they told stories about how they had experienced other people *using* the hegemonic discourse against them, assuming that they, as women, did not really understand the computer. And it is this unwillingly inscription in the hegemonic female subject position that made one of them exclaim that she was equal to men. There is – or ought to be – room for women, but women need to become proactive to claim this space. And they should not do it by trying to become like men, she claimed: 'We are not men. We don't think like men. But we have values that are just as good as men's, but we have to show that we have them.' Men and women are equals, but they are also different. Thus, the room this woman is trying to define includes both a masculine and a feminine subject position; men and women's relations to computers are still assumed to be different, but the masculine position is no longer the norm or defining 'proper' computer use. Because the hegemonic discourse was commonly used against her, her strategy involved a very active 'nagging' about her right to be heard, to perform, to be taken seriously, and to avoid being run over by boys who 'constantly try to steal the attention from girls'. Women have to be clear in their messages – they need to protest, because the feminine position, which should be equally as valuable as the masculine position, does not exist – it needs to be created. Also, she had experienced other people treating her according to the hegemonic discourse, but she claimed that through constant protest she was able to change the situation and to achieve an equal position. A study of women in academia in the US found that each new generation started out with the perception that gender equality had been achieved, only later to find this to be wrong (Bagilhole et al., 2008, p. 36). In a discussion between two of the students in Humanistic Informatics we can see indications of a similar situation. The youngest of them claimed that she was, or at least could be, equal to men, if only she demanded her rights. The other woman, a little older (whom we will meet next), rejected the young woman's optimism, insisting that an idea about masculine values as more valuable than feminine values permeates society in all aspects. The young woman's insistence on her equality also has similarities with how women in the UK have been found to insist that they work in a gender-neutral environment (Powell et al., 2006, p. 694), and to the 'strong belief in equal treatment of citizens' found in France (Godfroy-Genin, 2009, p. 85).

The positioning strategy illustrated here requires that the hegemonic discourse is actively rejected. However, the aim of this strategy is not to reject gender as a meaningful category, but to express a wish for women to be treated as men's equals. It is as women that they try to create a positive space for themselves. While keeping gender as a meaningful category, they reject gender as the basic category defining inclusion/ exclusion in relation to computers. The positioning strategy is, however, shaped as a reaction to situations in which gender *has* been allowed to provide negative associations between women and computers. It is with these particular situations in mind that one of them claims 'gender does not mean anything'. Thus, we can question whether their strategy is aimed at what they *want* gender to mean to a greater extent than reflecting their experience of what it means. The goal is a discursive room with equal space for masculine and feminine subject positions: not the deficient feminine position of the hegemonic discourse, and not a feminine position that is similar to the masculine, but one that is equally valuable. Although they reject the hegemonic discourse as a valid description of themselves and other women's relationship to computers, they do not reject its existence. Thus, this positioning strategy implies an idea of gender equality within computing, equality with room for both male and female values. Different from the previous positioning strategies among women is that these women do not consider themselves to be strangers in a masculine field – they have *not* entered a forbidden world, but claim their right to have a place, as women, within the world of computing.

A cross-gendered position

The last positioning strategy, involving a cross-gendered position, is only found among women. In other words – certainly not surprising, but still necessary to emphasize – none of the men shaped their positions based on similarities with the subject position associated with women.

Women doing men's stuff

Three women illustrate this last positioning strategy in which they claimed a position within the room for men. All three women were a little older than the average student, and none of them had any childhood experience with computers. They were, however, very enthusiastic and eager to learn, and expressed their fascination with – and even love for – the computer. Perhaps not coincidently, considering their age, these women expressed more strongly than the other women that society in general is permeated by men and masculinity as a norm, from religion to work environments: 'These thoughts reflect the structure of our society, where masculine values are always treated as better and more serious, even more human, than female values', the same argument used in conversation with one of the younger women who claimed to be equal with men. When these women claim their position within the room for men, they do it based on their previous experience of performing tasks most often performed by men. They refer to having grown up without a brother, having a father who took them along to work with tools, or being a single mother without a man to help, meaning that they had to do practical tasks, in particular involving tools and technology, associated with masculine subject positions. They call themselves 'untypical', and describe themselves as becoming 'more and more masculine' because they have to 'manage without a man', doing 'all the male things that women can manage'. They do not claim that they become men, but – to a certain extent – they act as men, and it is this particular experience that grants them access to the masculine room of the hegemonic discourse. In their study of women in IT work in the UK, Adam et al. found many examples of women attempting to hide their femininity, to become a genderless 'it', or to become more like men in order to be accepted in technology and engineering. 'It's the ideal career for women who have grown up with lots of brothers', claims one of the female engineers interviewed by Adam et al. (2005, p. 294). Although the women in Humanistic Informatics are not fighting an all-male environment as were the women interviewed by Adam et al., in this positioning strategy it is the exact opposite that is argued. It is primarily the absence of brothers or other male relatives that has provided them with the space necessary to acquire 'men's skills'. Thus, we can hear the echo of Oldenziel's claim that 'technology is what men do and women do not' (2001, p. 128), and when there are no men to 'do' technology, women might take over. This also points to the 'paradox' noted by Felizitas Sagebiel and Jennifer Dahmen, that single sex education, despite having a 'bad "odour"', might work to highlight differences between women, instead of differences between men and women (2006, p. 11).

The women referring to their experiences of doing men's work made no attempt to rewrite the discourse; the discourse remained intact with a positive masculine subject position involving computer knowledge, and a feminine position as a negation. They used the hegemonic discourse, and it is as women that they claimed their position in the room for men, though with a particular experience associated with men, almost as if they found being a woman in computing to be unlikely. And perhaps due to their general reflections about male dominance in society, the only possible way into a male-dominated computer culture appeared to be through the masculine subject position. Thus, gender remains intact, and men and women are still assumed to have very different relationships to the computer. Their advantage is that they have experience imitating masculine practices and using masculine tools and artefacts, which became their door key to the masculine position.

The missing positioning strategy: women's own room

Finally, there is one missing positioning strategy. When the students were asked how they perceived men and women's relationships to computers, the male students assumed that women had special skills in communication, design and aesthetics (that is, 'women and homosexual men'). These activities, as well as women's special abilities together, created a special feminized room in computing, according to the male students, a room where women were both assumed to have positive relationships with computers and where they were assumed to be better than (heterosexual) men, not unlike the feminized discourse we saw within recruitment initiatives in computing in Chapter 3. This room is, however, missing in the analysis of the female students. Although many of the women claim to like these activities, none of them claim that the most important foundation of their relationship with computers is based on special feminine qualities, women's ability to communicate or to be creative. Thus, the absence of these arguments and the lack of using these opportunities in positioning statements is probably not because women are unaware of or do not appreciate them. It is more likely that they remain insignificant in their positioning strategies because the assumed feminine specialities in relation to computing might signal something positive about women, but not in relation to the masculine norm of the hegemonic discourse of computers.

The power of discourse – freedom of individuals

The analysis of how computer students articulate different positioning strategies illustrates quite clearly how neither men nor women are homogeneous groups (Braidotti, 2002; Powell et al., 2004; Faulkner and Lie, 2007). Among both genders we find that they see a number of different possibilities, even when facing a masculine hegemonic discourse. All the positioning strategies use the hegemonic discourse as a valid frame of reference. It is not necessarily accepted, but it is seen as an existing discourse, as something they meet and have to deal with, and they deal with it in different ways. It is through 'what they do with what the world does to them' they construct their positioning strategies. Within the possibilities they articulate there are some gendered patterns which seem to open up or restrict their perceptions of themselves as computer users. We could probably find other subject positions and other positioning strategies among other social groups, in other contexts. The trend shows, however, that it is easier for men to be associated with computer-literacy. Based on gender, men can easily be ascribed a positive position relative to computers – for some men, too easily. Women on the other hand, have to negotiate in order to be ascribed a positive relation to computers. In this landscape, women are 'the others' – outside the masculine norm (Mahoney, 2001; Adam et al., 2005). To some of the women this becomes a shelter ('because I'm a woman and don't understand any of those [computer programming] things'), while it is more problematic for others, who raise a protest against being excluded. If we compare the strategies among men and women, we can see a greater variation among women's negotiations with the discourse than can be seen among men. Women introduce a greater number of new elements in their discursive negotiations than men, who seem rather to line up in a continuum according to how well they conform to the masculine subject position of the hegemonic discourse, without questioning the discourse itself. Although they propose new and alternative ways of interpreting gender and computers, only one of the strategies actually threatens the hegemonic discourse, which is the strategy aimed at achieving a common room in which gender exists, but does not define inclusion and exclusion. This strategy is the only active protest against, and therefore also threat to, the hegemonic discourse. The hegemonic discourse, despite being widespread and used against the women, is rejected as a mistaken and invalid description of the real situation, and they suggest replacing it with a different understanding of the relationship between gender and computers. Thus, this positioning strategy creates what Laclau and Mouffe call a 'social antagonism'; a situation where different discourses compete with each other and threaten each other's existence (Laclau and Mouffe, 1985, p. 125; Winther Jørgensen and Phillips, 1999, p. 60). In this case, it is the discourse of gender equality competing with the discourse of computers, and the former threatens to reject or dismiss the latter.

The different positioning strategies document that it is possible for the individual to negotiate what is perceived as hegemonic discourses and subject positions. In other words, it is possible to imagine that being a man or a woman in relation to computers can have other meanings than those of the hegemonic discourse; gender can mean something new (Annfelt, 1999, p. 370). By negotiating what they perceive as available subject positions they also contribute to the construction of gender, by proposing that being a man or a woman with a relationship to computers can have other meanings than those described by the hegemonic discourse. They become 'active subjects' (Søndergaard, 2000, p. 67). Not only can the women we have met here be *interpreted* as 'post-traditional actors' who have internalized the message about women's expanded room for actions (Annfelt, 1999, p. 77), they also *feel* that way. They might see themselves as 'typical girls' compared to boys, and as women who are different from men, but compared to other women, they are 'untraditional', 'new thinking' (cf. Powell et al., 2006), pushing gendered borders (cf. Mörtberg, 1997), and eventually entering the 'forbidden masculine world' of computing (cf. Adam et al., 2005).

Both in academic discourses and in everyday discourses we refer to myths about gender and computers. These myths have comprised a rather unclear area, often treated as unsettled questions or as 'fallacies' that can simply be rejected. What we see here is, however, that it is important to take such myths seriously – not as myths meaning something which is not true, but as cultural stories about the relation between gender and computers. 'People may or may not be consciously aware of their own biases, yet they impact the way in which people make decisions', Hayes points out (2010b, p. 271). As long as these cultural stories are perceived as a valid frame of reference to men and women who are trying to find their own positions in relation to computers, they also have real effects on real people.

Something stable that is constantly changing

As we have seen, the male and female students share an understanding of the hegemonic discourse about computers involving expectations of men having a closer and more 'gender authentic' relationship to computers than women have. But we have also seen adjustments and modifications, even rejections of the discourse, both in relation to content and in relation to where the discourse holds true. The discursive space for men changes relatively little, and the male positioning strategies are based rather on whether the men conform to or want to qualify for the masculine subject position. Only one of the men questions the hegemonic descriptions of men, not by questioning the discourse itself, but rather the degree to which the discourse has validity for all men. Among the women we see a more extensive negotiation of the discourse, both through the more open room for women (which, despite relying on the hegemonic discourse as a general frame of reference, also

to a larger degree acknowledges that women are engaged in computing) and through the common-room strategy (which is not primarily an
adjustment, but rather a rejection of the discourse). However, the hegemonic discourse exists and has real effects as long as other people base
their articulations or actions on it, and this also affects those women
who reject the hegemonic discourse. Both men and women, across the
positioning strategies, question *where* their alternative understanding
is valid, illustrating how some of the major modifications are not perceived as general, but are closely tied to the local context. Modifications
and adjustments do not abolish the hegemonic discourse – it still exists,
and it still affects both men and women. Neither do the modifications
and adjustments abolish gender, but still rely on gender as a difference,
although meaning different things. Several gender researchers have
pointed out how gender is at the same time both stable and changing:
stable because it always means a difference between men and women,
and changing because the exact content changes over time and place
(Søndergaard, 1996, p. 411). Hanne Haavind describes gender as a system
that is always divided in two (male and female) and always a hierarchy
with the male side of the dualism on top, albeit with changing content
(1994). Thus, we end up with an understanding of gender as two kinds
of 'something' that is constantly changing, and the only stability is
that the system – the dualistic split and the hierarchy – will always be
re-introduced, even when the meaning of gender is changing.

Is the variation we have seen here the beginning of a changing
discourse, or simply local variations? Individuals do not leave traces,
Søndergaard claims, and for a gendered field to change gender connota-
tion, it takes a number of opposite-gendered individuals over a longer
period (see Chapter 7). Only then will the gender connotations gradually start to change (1996, p. 193). This material alone cannot give an
answer to whether the individual modifications and adjustments we
have seen are symptoms of change. However, in Chapter 6 these findings will be discussed in a wider framework, together with the findings
from the other chapters. One thing can, however, be concluded from
this material; computers *can* be associated with new and alternative
gendered meanings. Gender-technology relations are not stable and
fixed for all time, but hold the potential for change.[7]

5
Stories about Individual Change and Transformation

Introduction

In Chapter 4 we saw how male and female students attending the programming course in Humanistic Informatics acknowledged the hegemonic discourse simultaneously as they constructed their own relations to computers and computing in different ways. We got to know the power of the discourse – a discourse that in certain ways was more inclusive to men than to women. We also saw the freedom of the individuals, which enabled us to see considerable variation in how both men and women define their relations to computers. In this chapter we will still follow this group of students, but this time we will focus on the female students and their stories about how their relations to the computer changed during the computer course.

The semester in Humanistic Informatics seems to have made some fundamental changes in many of the women's relationships with computers, and they all expressed a growing degree of self-confidence in that regard. However, what fascinated me the most when observing them in the computer lab, as well as in email correspondence and interviews, was their profound expression of *pleasure* in working with and learning about computers, and it is this pleasure we will explore in this chapter.

Is enthusiasm about computers a male thing?

Studies of women's experiences with computers have often focused on problems women encounter as a minority in a male-dominated arena, or on how women negotiate being both a woman and having a close relationship to computers, as we saw in Chapter 4 (Mörtberg,

1997; Kvaløy, 1999; Lagesen Berg, 2000; Langsether, 2001; Levold, 2001; Nordli, 2003; Adam et al., 2005; Powell et al., 2006; Godfroy-Genin, 2009; Bagilhole et al., 2008). It has been reported that women are not particularly interested in the computer itself. They are not interested in the technical aspects of computers or technical subjects (AAUW, 2000), often regarded as the most important subjects in computer education (Ahuja, 2002). Instead, women see the computer as a tool, and find its usefulness and its social aspects interesting (Håpnes, 1992). Early studies of women's relationships with the computer leave us with an impression that women use the computer because they find it useful or necessary, not because they want to or because they like it (Aune, 1996). Early studies of girls and ICTs have shown that they use the computer primarily for communication and obtaining information from the Internet, or for writing (Håpnes and Rasmussen, 2003). Studies of men, however, have often directed attention to men's fascination with technology and their absorption, and even love, for computers (Turkle, 1984; Hacker, 1989; Mellström, 1995; Kleif and Faulkner, 2003). One such male figure is the hacker (Levy, 1984), and one of the most famous descriptions of the hacker lifestyle comes from Weizenbaum's description of the 'compulsive programmers' from 1976, as 'bright young men of dishevelled appearance, often with sunken glowing eyes... oblivious to their bodies and to the world in which they move. They exist, at least when so engaged, only through and for computers' (1976, p. 116). This image has not changed much since 1976 (cf. Nordli, 2003), and the figure of the hacker has come to represent a typical male relationship with the computer. Men are not only perceived as skilled computer users, but also as being fascinated by and enthusiastic about the technology, in contrast to women (cf. Sørensen, 2002a). Gansmo et al., however, encourage us to 'forget the hacker', claiming that, despite the hacker being a minority among computer users, the image or stereotype of the hacker has had an impact on both research, politics and educational institutions (2003a, p. 36). Even so, it seems that enthusiasm about technology is mainly found among men, and while male enthusiasts have been an interesting topic to researchers, women's feelings for the technology is rarely discussed and, if so, mostly connected to dislikes, anxiety or reticence (Turkle, 1988). But is it true that women are not enthusiastic about computers? In this chapter that is exactly what we will prove to be wrong when we meet the women from the Humanistic Informatics course once more, but this time focusing on how they expressed joy and pleasure about computing.

Men's pleasure – a barrier for women?

The gender gap within computing has caused worries and has given rise to attempts to explain why women do not seem to want to enter the field of computing since the 1980s. It has been claimed that technology is linked to masculinity, that it is 'embedded' in a masculine culture, and that it acts as a signal of masculinity, and is thus perceived as incompatible with femininity (Turkle, 1988; Wajcman, 1991; Grint and Gill, 1995; Håpnes and Rasmussen, 2003). The associations between masculinity and technology, and computing as 'a masculine world' (Mahoney, 2001, p. 171), are among the oft-repeated explanations for women's absence from computing. Men's relationship to technology has often been referred to in terms of 'intimacy' and 'love' (Turkle, 1984; Hacker, 1989; Faulkner, 2005; Webster, 2005), and the special pleasure in technology found among men has also been identified as a barrier for women. In Aune's early 1990s interviews with families about their usage of computers in the home, some women claimed that it was their husband's intense and time consuming use of the computer that made them stay away (1996). Others have pointed to how women define themselves as that which men are not (Turkle, 1988; Mack, 2001; Håpnes and Rasmussen, 2003), of which we also saw examples in Chapter 4.

In their study of 'men's love affair with technology', Kleif and Faulkner claim that '[e]ngineering may be one of the most pleasurable of occupations' (2003, pp. 296–7). They found that men express a greater pleasure in technology than women. Men's satisfaction is explained by technology as a logical and rule-driven system that they feel they can control, in contrast to the impossibility of controlling unpredictable human beings. Turkle suggests that while women tend to be focused on people, men are more focused on objects. These men 'come to define themselves in terms of competence, skill, in terms of the things they can control' (Turkle, 1988, p. 44). Kleif and Faulkner argue that men's pleasure in technology reproduces male dominance within engineering, partly because the (assumed) male focus on technology stands out as incompatible with women's (assumed) interest in people. Men's intimacy with technology has been described as 'sensual absorption, spiritual connection, emotional comfort, and aesthetic, even erotic, pleasures' (2003, p. 297, cf. Hacker, 1989). Men's fascination and pleasure become signals of computing as a special masculine field, and in order to 'preserve' their femininity, girls and women reject the intimacy with the computer (Turkle, 1988; Aune, 1996; Håpnes and Rasmussen, 2003). Thus, as long as an intense relationship to technology is seen as a vital part of masculinity, a more distanced relationship with technology has come to define women (Wajcman, 1991).

The negative effect of men's pleasure has also been reported among women who have chosen to study computing. Margolis and Fisher found in their study at Carnegie Mellon University that women started their career as students in computer science with both interest and pleasure in working with computers. However, women discovered that the male students appeared to have greater knowledge than they did, and interpreted this as being 'more interested'. This made some of the women conclude that they 'were just not interested enough', and they lost their motivation to continue with computer education (2002). It seems that men's and boys' fascination with the technology makes women refrain from portraying their own relationship with the computer in terms of interest or pleasure.

Although several studies of women working in IT report that women express 'job satisfaction' in computing (Adam et al., 2005, p. 288), and that 'women engineers are...inspired and excited by engineering technologies' (Faulkner, 2005, p. 17), or that 'women can be strongly attracted to technological work' (Webster, 2005, p. 5), few studies have focused specifically on women's pleasure. A notable exception is Jane Abbate's study of women who had a career in computing, including some of the well-known pioneers, such as Anita Borg, Steve (or Stephanie) Shirley and Judy Clapp (Abbate, 2010). Abbate's very interesting interview material profoundly illustrates how these women, whose careers began in the 1950s, 1960s and 1970s, loved what they were doing. They express deep fascination not only for the technology itself, but also for computing being a field suitable for women, as one of the pioneers illustrates: 'I still think, of all the fields open to women, computer science is the most wonderful one. First of all, as a programmer, no one knows what sex you are, what color you are, what your gender preferences are; they just know: Does it work or not? Did you get it done? Is it fast enough? And therefore, it is *the* field where you are judged by the output – that's it...So I love it for women' (Abbate, 2010, p. 213). Thus, women's fascination for technology might have been discursively invisible, but it has definitely not been non-existent.

Also, a study from Norway explores women as computer enthusiasts, as Nordli wanted to find female hackers (2003). She did not find women who agreed to call themselves hackers, but she did find female computer enthusiasts. She found 'professionals', 'IRC-babes' and 'Geek-grrls', women who have more or less dedicated their lives to working and playing with computers. Nordli's study demonstrates that female computer enthusiasts do indeed exist, but it also demonstrates how these women have to negotiate the dominant expectation that women do not have an interest in computing or are not computer-literate. Some of the women who participated in a group of computer enthusiasts even expressed this expectation themselves, worrying that new women entering the group were 'stereotypical' girls who were not really interested in computers, and therefore could undermine their own position in the group (Nordli, 2003). In other words, they did not expect to find other women with the same intense relationship to the computer that they experienced themselves.

Surveys of study motivation among students in computing or engineering also document women's interest in the field. A quantitative study of motivation among students at a computer course at the Norwegian University of Science and Technology (NTNU) in the late 1990s found that both men and women enjoyed the course when they felt that they could master the assignments and the technical demands (Lagesen Berg and Kvaløy, 1998). It is perhaps no surprise that mastery increases the comfort and feeling of wellbeing, as is the case for the women we will meet again shortly. A study of student enthusiasm for engineering in the UK shows perhaps more surprising results, as it found that men's and women's self-motivation decreased over the first two years, but subsequently women's self-motivation increased once more, while men's self-motivation continued to wane (Alpay et al., 2008, pp. 580–581). This study also showed that women scored high on 'enjoyment of maths and physics' as their specific reason for choosing engineering (ibid., p. 578). Thus, while this study does not specifically address women's pleasure in technology, it documents that women might be highly motivated to achieve technological careers, partly for reasons that have previously been assumed to exclude women (cf. Hayes, 2010b). In spheres outside education, a study of girls' computer use in Norway found that teenage girls took pleasure in and became fascinated by Internet services, as well as computer games. However, it was also found that girls socialize each other into suppressing expressions of pleasure in technology (Håpnes and Rasmussen, 2003). As long as technology is perceived as a masculine field, fascination for technology is also discursively reserved for men, and 'girls learn to adapt to what they think is the prevalent female image and to avoid behaviours that deviate from these images', Sagebiel and Dahmen conclude (2006, p. 11). Also, due to the masculine culture around technology and men's fascination for technology, women have not always found the contexts for cultivation of such fascination inviting. Computer games have been one of the important fields to foster fascination for technology, and Raina Lee, the producer of a 'megazine' for conveying 'computer games pleasures' for adults, including 'feminist readings of games', explains her motivation with reference to traditional game material targeting young boys: 'Most of the material comes from game publications that are written in the voice of a 14-year old boy, which is ok if you are a 14-year old boy. Many of us are not, and have never been' (Dovey and Kennedy, 2007, p. 24).

In the following we will explore how the female students[1] from Humanistic Informatics express pleasure in relation to computers and computing. As we recall from Chapter 4, they were students at a first-term bachelor's course in computer programming, and even though all of them had used a computer before, most of them described themselves as inexperienced. Thus, they were not 'professionals', 'geek-grrls' or 'IRC-babes' (Nordli, 2003) when they entered the course. Their motivation for attending the course varied, from being curious, to not wanting to be 'illiterate' any longer, to feeling that they needed computer knowledge. One started because her girlfriend had started, while others took the course while waiting for admission to other courses. They did not spend all their leisure time working or playing with computers, and they did not plan a career within computing.

In Chapter 4 we explored how expectations towards men's and women's relations to computers create certain gendered positions, and how the male and female students related to these positions when they created their own relationships with computers. We saw that all the students supported one particular understanding of gender and computing, a dominating or hegemonic discourse, in which men are expected to have greater interest, experience and knowledge about computers than women. Men are expected to be fascinated by the technology, and they are associated with computer games, programming and technical tasks, while women are associated with tasks such as communication, information and writing. In short, women are, discursively, 'the others' in relation to the masculine associations of computer technology. What puzzled me was that, contrary to what we could expect, the female students expressed great pleasure in their relationships with computers.

Women's pleasure in computing

Most of the women expressed pleasure in a wide range of computer activities, and some were thrilled that they had finally found the courage to attend a computer course. The pleasure they expressed is perhaps not surprising, since they had all chosen to work with computers.

Still, it is not an obvious reaction, and there were a few women who expressed more dissatisfaction than pleasure, in particular the women emphasizing their harmony with the hegemonic discourse; those who were dissatisfied because it was too much, too difficult and too uninteresting. But these were the exceptions rather than the rule among the 21 women. Shouts of joy occurred more or less daily in the computer lab, and the sounds gave clear evidence of pleasure in working with computers: 'It was great fun when the code passed through and I achieved exactly what I had been working on for so long... Oh yes, great jubilation in the computer lab then...' (Ingunn). As previously mentioned, the connection between mastery and pleasure was also evident among these students, and there were no obvious gendered differences in this respect. However, the cheers of joy were much more apparent among the women than among the men. It was mostly women who drew attention to themselves in the computer lab with their laughter, or made the whole class gather around them, showing them some funny things they had created on the computer. But how could these 'others' (women) express so much pleasure in relation to something that was not 'theirs'?

In the following, we will explore how the women articulated their pleasure in computing, first, by looking at what it meant to be a woman on a computer course. Second, we will explore what it meant to these women to attend the course, before we turn to the activities they found especially pleasurable.

The unexpected female computer students

Women are not expected to be found in large numbers on a computer course, and many of the women were surprised to find a majority of female students in Humanistic Informatics.[2] They saw it as evidence of a 'women want, women can' attitude among women, and claimed that it improved their self-confidence. The female lecturers were also seen as positive role models: 'I think that it is very positive that there are female lecturers. It gives inspiration and it helps with self-esteem. When a lie is continually repeated, you believe it ("Girls are lousy with computers")' (Marte), one of the women wrote to me in an email. The presence of women helped to repudiate the 'lie'. It also affected their perception of computing: 'Earlier I had the perception that computer culture was masculine, but I have changed my opinion on this, perhaps because of the large majority of women in this course' (Turid). The fact that female students were in the majority was an important factor in enabling the women not to see themselves as 'misplaced': 'the girls encourage each other; you see that the choice [of education] is not quite as strange for a girl, rather the opposite; here, the boys are in a clear minority' (Bente). It seems that the female majority had a positive effect on the women's wellbeing as computer students. Once more, the exceptions were the dissatisfied women in harmony with the hegemonic discourse. To them, the majority of female students was less important than the way they found the male students dominated the classes, and the female lecturer did not become a positive role model, but was seen rather as having crossed the gendered border and become absorbed by the masculine culture of computing.

Even though women are not expected to be found in large numbers in computing, *being* a woman in a computer course was not necessarily seen in a negative light. Some of the women even thought that they had certain advantages because they were women: '[T]he expectations of a girl who is doing things with a computer are not as great as those for boys....the people are more impressed if a girl manages something in the field of computing' (Kathrine). They felt that they received more positive attention than boys and men working with computers, which has also been found in studies of women in engineering in the UK (Powell et al., 2004, 2006). This is clearly a double-edged sword; it can be seen as well-meaning and positive encouragement, however, it also points to the low expectations people generally have of women's computer skills.

What does it mean to attend a computer course?

The female majority might have made it easier for these women, but what did it actually mean to them to be computer students? 'It is of incredible significance to me to have started on a course like this, since I lacked all the basic knowledge, I didn't know anything! Now I can do thousands of things that I couldn't do before' (Mette). Mette felt that she was 'finally doing something useful', and after only a few weeks, she felt more confident and had learnt a great deal. However, this surprised her: 'This term is the best term I have had at the university. It exceeded all my expectations, [I] thought it would be more "incomprehensible", if you see what I mean, but it was actually possible for me to absorb the knowledge, in spite of a lousy starting point.' Even though the computer was not new to Mette, she described her starting point as 'lousy', and she was surprised, not only that she actually managed to learn something, but also about her new and intimate relation to the computer, and she claimed that it had 'become a tool that I am addicted to! You would never have believed that about me!' The female students quite clearly see how their own relationship to computers has changed, or even transformed in unexpected ways. They not only act in new ways previously unexpected, they also *feel* in ways they had not expected.

One of the most sceptical women, Lillian, was also surprised by her new knowledge. She was one of the women using gender to explain her lack of skill and understanding (Chapter 4), claiming 'I don't understand any of that [computer programming], because I'm a woman!' and also strongly resistant to learning the 'engineer stuff': 'I am not going to be an engineer. ... I don't understand anything ... I don't want to understand it either. I don't want to learn it!' She was discontent, striving hard to learn, and claiming that it is natural that women understand less about the computer than men. Lillian did, however, learn something at the course, and also she expressed pleasure about her new knowledge. She told about how her husband, who 'didn't have a clue' about computers, kept 'messing things up' – the same way she previously had, but now she could fix it. Lillian's pleasure was quite obvious when she exclaimed happily: 'I know something! I know something!' She was excited – not about any particular things she had learnt, but about how she actually had learnt something about computers.

Another woman, one of the slightly older women justifying her place in computing by referring to her experience of performing practical tasks typically undertaken by men, fulfilled a dream by attending the course: 'I have always wanted to learn a little bit more about the computer. I have sort of seen that as very unattainable, but when I got in here [on the course] it became attainable' (Tone). She was quite new to the computer and she worked hard to understand it, probably harder than most of the students. Yet, she maintained her positive attitude and was grateful because she finally learnt more about computers.

Several of the women compared computer knowledge to literacy: 'To me it was a conscious decision to enrol in a computer class. I did not want to continue being the illiterate that I believed myself to be', Marte claimed. Being literate is not only a matter of being able to read or write – or in this case, being able to use the computer, it also refers to being included or excluded from important arenas and activities in society. Thus, to Marte, the computer course meant that she was about to acquire skills that appear to be necessary in our society. Another woman pointed to how her new computer skills affected her social position in another way: 'We haven't had much practical work with hardware, but we have acquired a little theory, and now I feel that I am more prepared when men talk about RAM and processors etc. I am really happy that I took this course' (Anne). Also, Anne claimed that she had become better 'socially equipped'; she had become able to participate in men's discussions about hardware.

While Anne expressed her pleasure in relation to a male community, Helga expressed her positive experiences in relation to other women: 'I have advantages because I'm considered to be "fresh thinking" and courageous in choosing subject... a subject where you sort of depart from women's traditional areas, and at the same time go into one of the most important areas of the future.' Helga felt that she benefited from being a 'non-traditional' woman by attending a computer course. To move beyond fields that women traditionally have been engaged in is part of the postmodern project, according to Annfelt, and by doing so, women signal that they are 'post-traditional actors' who have internalized the message about an expanded 'room for manoeuvre' for women (1999, p. 77).

Helga had made an effort both to learn and to like working with computers before, but she had not been 'turned on': 'I started at the bottom when it comes to computer knowledge – but I feel that with every new day I master new things... It feels like a new world has opened up to me... and every day I think "How on earth is it possible to cope without knowing what I know today!??" It has to be a feeling close to something like going from being illiterate to being able to read...' Helga had used a computer before, and had even attended courses in order to bring herself 'up-to-date', but even so, she described her starting point at the beginning of the term as 'at the bottom'. Also, she compared computer knowledge to literacy and described how it felt as though she was entering a new world filled with indispensable knowledge, and in an email to me she exclaimed: 'I think that I have become addicted to the computer!!! Earlier I did not think it was either appealing or attractive – but now I can hardly manage a day without it... I use it as a tool both for learning and in daily life. It has become an important source of pleasure... [I] feel very different about it now, after having learnt how it works. It is not something distant and difficult, but a useful tool – and not the least a thing that I can manipulate and use as I want to. I am addicted – but I have the power over it – not the reverse...'

Helga experienced a 'revolution' in relation to the computer, and it had become a fascinating tool for both work and pleasure. Contrary to what we have seen in other studies (Turkle, 1988; Håpnes and Rasmussen, 2003; Sagebiel and Dahmen, 2006), she had no reservations about calling herself 'addicted'. Also, other women had experienced a 'revolution' in relation to the computer, for example Åse, who did not really like computers: 'I have got a much better relationship to the computer now. It has in fact come so far that I miss it after one day away from it. Before, I hated everything to do with computers. I couldn't understand that anyone would sit down in front of the machine voluntarily. I think differently now. I have in fact become fond of the machine' (Åse). Åse had started on the course because her girlfriend had started and, although she was sceptical at the start, she had 'in fact' – as if it was highly unlikely – become fond of the computer.

Sara had her epiphany a couple of years before she started on the computer course. Before this point, she saw computers as 'nonsense and rubbish' and she strongly rejected the idea that she would ever use a computer. Her first contact with a computer changed this completely: 'I had my first meeting with computers when I attended [another course in the late 1990s]. A friend of mine introduced me to a small Mac which I borrowed. I used it writing assignments at the course. I fell in love with the machine and dreamt about it at night and looked forward to the next session. Even today I miss the printer's beautiful melody and its rhythm when it drew in a sheet of paper and printed on it' (Sara). In spite of her previous negative attitude, Sara experienced the first meeting with a computer as so intense that she described it as a love story. Her fascination seems to have been directed at the beauty of the technology itself. One of her motivations for attending the computer course was that she wanted to learn more about how to use a computer in a creative way, and she had fantasies about how she could use the computer, especially programming, as a creative tool. She described her experience with the computer almost as a voyage of discovery and compared it to a craft: 'It has similarities to a craft, right – the medium itself wants something – you don't have complete freedom, but you can do something with it if you know how to do it' (Sara). In Sara's terms, the computer had become something more than 'technique and logic'. She both enjoyed the intimate relationship with the machine as well as the creative aspect, just as had been found among hackers and other male computer enthusiasts (Turkle, 1988; Hacker, 1989; Aune, 1996; Håpnes, 1996; Kleif and Faulkner, 2003). However, contrary to the pleasure found among male enthusiasts, it was not the controllable and predictable aspects of the technology that fascinated Sara, but rather the unpredictable – the things she did not yet know, which could, because they were still unknown, hide inconceivable possibilities.

As we can see, women express a variety of reasons as to how joining a computer course meant a great deal to them. According to the hegemonic discourse, women are neither expected to be interested, nor have experience or computer skills, and many of them were surprised that they could manage it and that they actually found it interesting. However, when they reached this stage, they were totally absorbed and 'addicted'.

The women we have met here expressed pleasure, satisfaction and pride in having attended a computer course. But what is it about the computer that these women like, and what are their favourite activities?

What do women actually like?

Girls like to use the Internet for communication and information and, according to earlier research, they like to use the computer for writing (cf. Håpnes and Rasmussen, 2003). Also, the students, both men and women, described these things as typical activities for girls and women. None of them, however, claimed that they enjoyed using the computer for writing, but a few of the women pointed to the Internet as interesting: 'It's fun to see the advantages of using the computer as communication – news groups and chatting – this is new to me!' (Helga). Interest in the Internet was also expressed by the men in this study, and it is not possible to see a particular gender difference here. There are, however, other aspects and activities that the women described as exciting and interesting. A few of them spoke of how they had been hooked on computer games and used games as relaxation in between other activities (Lise, Turid). Women do play computer games and they enjoy it (cf. Nordli, 2003). However, to some women it seems important to express their pleasure as moderate compared to men's pleasure (Håpnes and Rasmussen, 2003), such as the women who laughed and made fun of my question about what they did to have fun with the computer. As we continued talking about things such as computer games in this interview group, the three women agreed that games were fun, but they were also a 'boy-thing'. The three women admitted using games and that they found them enjoyable, but they also emphasized that they did not spend as much time or money as boys playing games. They made it a 'typical girl' thing to have a moderate experience of enjoyment in contrast to boys.[3] It seems that they relate to the hegemonic discourse and the different pleasures it makes 'available' to men and women; these women might play, but they described themselves as less fascinated than men.

Not all of the women played games, but they found pleasure in things they worked with on the computer course, including activities associated with men in the hegemonic discourse. To some of the women it was precisely this new and unfamiliar knowledge which seemed most attractive: 'The fun part is when you manage something which earlier seemed to be "totally inconceivable", such as the HTML code and programming. You wait in eager anticipation when you save it and check if you get it right, have to put in a great deal of effort to make the program work, until you get sick and tired of it, but you sort of can't give up, and when you achieve your aim, there is only joy and happiness ... There are perhaps more downs than ups, but the positive moments count twice as much!!!' (Mette). To Mette, programming had appeared as 'totally inconceivable' before, and she emphasized that the positive experiences overshadowed the hard slog. Kathrine also described her pleasure in finally achieving something: 'on that programming assignment, when you sort of understood more and then it was like "hallelujah" ... I wouldn't have believed I could make it, but ... I could' (Kathrine). As previously mentioned, pleasure and mastery go hand in hand, and many of the cheers of jubilation in the computer lab were undoubtedly connected to these moments. What is special here is, however, that many of the women expressed pleasure in relation to things they had not expected to master.

Many of the women also expressed a special pleasure when they talked about having knowledge that others did not expect them to have. Programming was described as one of the activities most exclusively associated with men: 'Maybe that is why I want to work with programming, because it is so masculine ... I feel sort of as if I were in a world that's a little bit forbidden. ... Very few expect that a woman can do programming. That is probably why I find it especially exciting. ... I think there is some kind of status symbol connected to it' (Bente). Bente was fascinated by the forbidden world of programming – the world where women do not have an obvious position, and she also suggested that she gained authority from having knowledge in a male-dominated field. Another woman spoke of how she enjoyed having the skills that impressed her girlfriends: 'I think it is much cooler to come and fix the computer [compared to programming]. A girlfriend of mine was pretty impressed yesterday when I defragmented her hard drive' (Marte). The person fixing a computer is normally a man, according to Marte, and she enjoyed having acquired these skills herself, because she could both surprise and impress others, even with quite simple tasks such as defragmenting a hard drive.

As discussed in Chapter 1, it has been claimed that technology is something men can use to signify masculinity (Lie, 2003b, p. 259), while '[w]omen's identity is not enhanced by their use of machines' (Wajcman, 1991, p. 89). Quite contrary to Wajcman's claim, it is clear that for the women we have met here, their new relationship with the computer did affect their identities, by providing them with 'status', or changing them into 'non-traditional' and 'fresh thinking' women (cf. Powell et al., 2006), and by giving them access to social contexts, such as men's techno-talk, in which they had not previously been able to participate. Thus, it is clear that women's identity *can* be enhanced by their use of machines!

Dissolved discursive understanding

If we see the women's expressions of pleasure in relation to the hegemonic discourse, we can, first, see that they crossed a gendered border by being engaged with computers. According to the discourse, they are 'the others', about to find their way into the new, exciting and 'forbidden' field of computing. Second, the women were surprised that they actually had acquired knowledge that they neither thought they would enjoy nor be able to master. Third, it seems as though the women felt that they gained access to some of the status that they associated with computing as a masculine field – especially the most 'exclusive' masculine areas, such as programming and the technical side of the computer. And, last, it seems as though the women used their new knowledge and its status to both surprise and impress others.

Their pleasure can be seen as a result of challenging or even dissolving their previous discursive understanding of computers. Through the computer course they experienced that computers are not incomprehensible or boring for women but, on the contrary, both fun and interesting. They discovered that they could cope and they could learn. And when they finally realized this, they enjoyed working with computers, almost as if they were 'starved' of computer knowledge. Cockburn has documented how men's exclusion of women, through monopolizing technical know-how, has been a strategy to preserve male dominance in technological contexts (1988 [1985]). In relation to computers during the period of this study, it is not men's (intentional) monopolizing which excludes women, but rather a hegemonic discourse that women have internalized, which gives guidelines for understanding gender and computers. As Hayes has pointed out, cultural images, whether we are consciously aware of them or not, still affect our choices (2010b, p. 271). If we had also included the male students in this search for pleasure in computing, we would have seen that the women's articulations about pleasure were much more expressive than the men's. We would also have seen that the men did not articulate pleasure about having 'discovered' that the computer is interesting after all. Neither did men emphasize gendered role models, or the importance of impressing friends. Also, the men expressed delight in having knowledge in a masculine field – not, however, as a 'forbidden' field, but rather as a more original masculine field that they seemed to feel they should have been familiar with in the first place, as a more 'gender authentic' field for men (Faulkner, 2005, 2006).

Gender is not the only factor affecting individuals' articulations of pleasure, but it clearly influenced how these men and women expressed their pleasure in computing. The reported incompatibility between women and technology might have had the result that girls and women's pleasure in technology is under-represented (Håpnes and Rasmussen, 2003; Sagebiel and Dahmen, 2006). Women might enjoy playing with computers, but some women moderate how they express their pleasure compared to men's pleasure when they talk about their own relationship with the computer. However, as we have seen, this is clearly changing, and a number of women in this study were not afraid to articulate their pleasure, joy and addiction, or even love for the computer. Thus, when we look at how the women expressed their pleasures in relation to the computer, we do not see an image of reticent women, as we could have expected from earlier research. Instead, we have seen women ready to express an intimate relationship with computers. It is evident that, even if few women participate in men's technological communities, as Kleif and Faulkner claim (2003), there is no reason to underestimate women's fascination and pleasure in learning more about computers. The fact that women do enjoy working with computers – and express it – could mean that the 'gendered code' of computers is about to change (Annfelt, 1999, p. 370).

We have met women with different motivations for attending a computer course, and for quite a few of them it seems to have been more a coincidence than a deliberate plan to achieve an intimate relationship with the computer. One of the challenges these women faced, and one which is still a challenge, is that stories about pleasure in computing do not seem to 'stick' to women; there are not many 'cultural stories' about women's pleasure in computing readily available for women to identify with. Instead, women meet cultural stories about computing and technological careers being unsuitable or 'gender-inauthentic' for women (Faulkner, 2006; Godfroy-Genin and Pinault, 2006). Perhaps one of the informants is right in saying: 'I think that some [women] have to be tempted to discover that it can be fun working with [computers]' (Lise). However, by inviting women into computer education because they are expected to be good at 'something else', which we saw in Chapter 3, confirms the discursive description of women as 'not really interested in computers' (cf. Corneliussen, 2003a; Nordli, 2003).

The women we have met here might not represent women in general, but they do portray real stories about women's pleasure in computing – stories that hopefully can be available to future female computer enthusiasts as cultural stories. We have seen here that women enjoy working with computers, and they are not afraid to express this, but they seem to do so only after having 'seen through' or exposed the delusion of their previous assumptions, or the hegemonic discourse, about gender and computer. And although these women operated in an environment where women represented the majority, here too the impact of the cultural discourses in which computing belongs to a masculine domain continued to play a role.

One of the concerns we have within computing is to change the hegemonic discourse so that it reflects the existence of women's joy and pleasure in computing, and their enthusiasm about the technology, not only about 'social and communicative' aspects, as we saw in Chapter 3. This task is not easily 'imposed' upon anyone, and as we saw in Chapter 2, the press has been rather resistant to communicating stories of this kind. However, we also saw signs of change, as cultural stories of women as computer users are becoming more typical. If we were to learn something from the stories of the women we have met here, it would, however, be that we need to acknowledge women's positive relationships to computers. And perhaps that could become the foundation for future inclusion strategies, thus inviting women to computer education by telling them that 'you might even fall in love with the technology!'

6
Layered Meanings and Differences Within

Introduction

Throughout this book we have explored gender-technology relations, with a special focus on how men's and women's relationships to computer technology are discursively described, perceived and dealt with in real life. Chapters 2 to 5 all focus on the same topic, but in different contexts and through different empirical material: Chapter 2 explores discourses in a newspaper and a computer magazine, Chapter 3 analyses discursive developments in higher computer education, and Chapter 4 is based on interviews and observations among computer students, and analyses their perceptions of the discourse and their own identity constructions involving gender and a gendered field of study. Finally, in Chapter 5 we explored women's fascination for and pleasure in computing. We have seen a number of examples of how gender and technology are woven together, in many ways, on many levels, in discourses and in lived life. However, there are still many questions to answer, such as, what does it mean that women have to give up part of their femininity to engage with computers? Or could that perhaps be an example of an academic 'truism', similar to the truism of the 'cultural association between masculinity and technology' pointed out by Grint and Gill (1995, p. 3)? Why, instead, do women not introduce femininity? And what does the discursive ignorance of women's contributions in computer related contexts mean? What kind of gender are women 'doing' (West and Zimmerman, 1987) in computing? And what is being 'undone' (Butler, 2004)? What makes a difference, and what matters to gender-technology relations?

Each of Chapters 2 to 5 focuses upon a particular context and a particular way of constructing gender-technology relations. However, in this chapter we will explore these contexts together, to investigate the constructions of meaning related to gender and computer technology from different angles, as the four previous chapters can also be seen as four partial perspectives (Haraway, 1991c). The aim of this chapter is to look for constructions and configurations of gender and technology across the discursive contexts, cultural stories and lived life, to explore connections that were not apparent in the previous 'near-sighted' chapters. Doing so, I will apply an analytical approach inspired by Karen Barad's 'diffractive methodology'.

By 'diffractive reading', Barad refers to a method of reading several texts 'through' each other, as 'a way of attending to entanglements in reading important insights and approaches through one another' (Barad, 2007, p. 30). There is a long tradition of using optical metaphors in research (ibid, p. 29), but instead of merely using the metaphor of 'reflection', Donna Haraway proposes the metaphor of diffraction (Haraway, 1992, cited in ibid.). While reflection simply reflects or makes a mirror image, diffraction is a physical phenomenon tied to the behaviour of waves, and 'diffraction patterns' are those new patterns arising when waves hit an obstruction or when they meet and combine (Barad, 2007, pp. 74–7). While reflection 'invites the illusion of essential, fixed position' the metaphor of diffraction 'trains us to more subtle vision' (Haraway, 1992, quoted in Barad, 2007 p. 29). With one foot in quantum physics and one in feminist science studies, Barad develops this idea, and claims that 'a diffractive methodology is respectful of the entanglement of ideas and other materials in ways that reflexive methodologies are not' (2007, p. 29). Research should not start with pre-defined objects, subjects, definitions and boundaries, Barad claims, but instead employ a method that 'enables genealogical analyses of how boundaries are produced rather than presuming sets of well-worn binaries in advance' (ibid., pp. 29–30). Her diffractive methodology starts from the premise that 'diffraction does not fix what is the object and what is the subject in advance, and so, unlike methods of reading one text or set of ideas against another where one set serves as the fixed frame of reference, diffraction involves reading insights through one another in ways that help illuminate differences as they emerge: how different differences get made, what gets excluded, and how those exclusions matter' (ibid., p. 30).

Thus, the diffractive methodology Barad proposes means reading different texts through each other, using a framework which does not take as a starting point that entities, objects and subjects have pre-defined boundaries, but instead urges us to pay attention to how entities, identities and boundaries are created in specific contexts. I will employ diffractive reading as an attempt to explore patterns across the studies presented in Chapters 2 to 5, and more specifically, to explore what diffractive reading can reveal about stability and change in gender-technology relations, which is a central question not only in this book, but also at the core of the efforts made since the 1990s to improve women's situation in computing.

A theoretical frame for diffractive reading

Referring to strategies to include women in computer science and tech-nological contexts, Maass et al. point out that 'the realization that men and women fall into more than two categories means that there is no single strategy that will work to include "all women"' (2007, p. 16). The same realization also means that we might have to explore this phe-nomenon with the help of several theories in order to understand how gender and ICTs are tied together in a seamless web of social, cultural and technological factors (Hughes, 1986), how they are co-constructed (Cockburn and Dilic, 1994b; Faulkner, 2001), and not least how resources of men and women, masculinities and femininities are unevenly dis-tributed in this seamless web. Thus, before we move on to the diffrac-tive reading, I will recap (see Chapter 1) and refine the theoretical frame through a set of theories that have been particularly helpful in the dif-fractive reading of the previous chapters.

The main perspective throughout this book has been tied to dis-courses; the importance of cultural perceptions, individual's actions, ideas and their co-existence with, dependence on and effects upon their environment. Discourse theory blends the focus on constructions of meaning with individuals' agency – the freedom of individuals with the power of the discourse. 'Discourse is not a synonym for language', Barad points out, but 'that which constrains and enables what can be said', and discursive practices are that which 'define what counts as meaningful statements' (2007, p. 146). The other central perspective has been tied to gender. 'Gender is not exactly what one "is" nor is it precisely what one "has"', Butler writes (2004, p. 42). What makes us men and women depends on what we do with 'what the world does to us' (Simone de Beauvoir, see Chapter 1), or, as formulated by Butler: 'If my doing is dependent on what is done to me or, rather, the ways in which I am done by norms, then the possibility of my persistence as an "I" depends upon my being able to do something with what is done to me' (ibid., p. 3). Important to this perspective is the notion of gender as some kind of 'doing', as emphasized by West and Zimmerman in their article

'Doing Gender' (1987).[1] In West and Zimmerman's perspective, '[d]oing gender involves a complex of socially guided perceptual, interactional, and micropolitical activities that cast particular pursuits as expressions of masculine and feminine "natures"'. Although it is individuals who 'do' gender, it is also 'a situated doing, carried out in the virtual or real presence of others' (ibid., p. 126). 'Doing gender' is not something we do in an empty space, but in context-bound social situations, or in the frame of a living society: 'participants in interaction organize their various and manifold activities to reflect or express gender, and they are disposed to perceive the behaviour of others in a similar light' (ibid., p. 127). This perspective stresses the importance of focusing on how discourses and agency interact, mutually influence each other, how we are shaped by and are shaping discourses (Søndergaard, 1999), as well as how we produce each others' conditions (Haavind, 1994).

As with Madeleine Akrich's description of script theory (1992), pointing out that despite technologists putting 'scripts' into artefacts, users are free to alter or ignore the scripts, there are (many) gender scripts – norms and guidelines – but the individual is free to change them or ignore them altogether. The post-traditional society, Giddens claims, has increased our opportunities to choose our lives – we are no longer tied by traditions, but the individual is also always interacting with and dependent on the larger structure (Giddens, 1984, cited by Gauntlett, 2008, pp. 101–2). The post-traditional freedom does not, however, remove the need to remain recognizable as men and women to avoid rejection or reprisals (Søndergaard, 1996, p. 33).

The notion of gender as 'doing' has been re-phrased, re-worked and developed in various ways. Judith Butler's theory of gender as performative is often used to emphasize that gender is not an essence we are born with, but rather something constantly in the making (1990). 'Doing gender' can be quite concrete, Butler reminds us, describing men at a drag bar who 'could do femininity much better than [she] ever could, ever wanted to, ever would' (2004, p. 213). Thus, central to Butler's theory about gender as performativity is the repetition of gendered behaviour, patterns, norms and so on, but it is also 'always "doing" with or for another, even if the other is only imaginary' (ibid., p. 1). In her book *Undoing Gender* (2004), Butler explores the situation of becoming undone and unrecognizable. Still based on the notion of gender as performative, Butler explores expressions of gender that are unfaithful to the 'matrix of the '"masculine"' and '"feminine"', pointing out that 'the production of that coherent [masculine and feminine] binary is contingent, that it comes at a cost, and that those permutations of gender which do not fit the binary are as much a part of gender as its most normative instance' (ibid., p. 42). Thus, the notion of 'undoing gender' refers to processes of transforming, remoulding, rejecting or in other ways being unfaithful to gender norms, resulting in gender expressions that often remain concealed in the shadow of hegemonic or stereotyped narratives. Or when revealed, expressions that might cause negative reactions. Through her discussions of expressions of sexuality outside the heterosexual norms, Butler encounters far more violent, even life-threatening, reactions towards the unfaithful permutations than we find when discussing gender-technology relations. However, the notion of undoing gender as a 'critical and transformative relation' to norms might be helpful in exploring men's and women's experiences with computer technology as well as perceptions of men and women, masculinities and femininities, in relation to ICTs. The kind of undoing I will illustrate is particularly effective because in most cases it is subtle, and within the limits of what we find to be wrong, but still acceptable (or perhaps unavoidable). Hege Skjeie and Mari Teigen explain a similar attitude in relation to gender equality policy; as long as we believe that the situation is gradually improving, we tend to accept continuous inequality (2003).

While the 'doing' gender perspective is aimed at explaining how gender is produced, maintained, reintroduced and reinforced, the 'undoing gender' perspective aims to explore those inharmonious expressions of gender that do not conform to heteronormative social norms. In short, the first aims at explaining how gender is made to be what it is, the second aims at explaining what happens when gender is not what it ought to be. 'Undoing' needs the normative to be able to 'undo' it, creating a paradox in which transformative forces rely on conservative figurations. However, at the heart of the concept of 'undoing' we also find a potential for creating change through a reworking of expected or normative figurations.

In the following I will not only refer to gender as being 'undone', but also to the doing and undoing of 'technicity'. The concept of technicity has been used by Gilbert Simondon to refer to technological qualities of technological objects or technological dynamics of technological objects (1980 [1958]). The concept of technicity has also been used by David Tomas (2000) to describe the creation of identities and groups in the cyborg-worlds described by William Gibson in his novels 'to describe ethnic-type relations among cyborgs, especially since traditional blood ties are increasingly replaced, in threshold cyborg cultures, by technologically defined social bonds' (pp. 184–5). Thus, Tomas uses the concept as an alternative to 'ethnicity' in a situation where technology enters group- and identity-formations in vital ways. From Tomas, the concept of technicity is taken up by Jon Dovey and Helen Kennedy as a way to 'account for particular formations of identity and power which lie at the heart of computer game cultures' (2006, p. 18). They emphasize the concept's ability to 'encapsulate, in conceptual terms, the connections between an identity based on certain types of attitude, practices, preferences and so on and the importance of technology as a critical aspect of the construction of that identity' (Dovey and Kennedy, 2006, pp. 16–17). They also illustrate how creativity, technical interests and abilities produce specific expressions of identity in which technology is a vital part (Dovey and Kennedy 2007). It is in line with Dovey and Kennedy's interpretation that the concept of technicity will be used here, referring to how technology, technological skills, preferences and images are involved in identity construction. Technicities 'are never fixed, never completely determined, but are contested and negotiated, technicities are a "becoming" in themselves produced through our daily encounters with technology and our ever shifting tastes, desires, pleasures and competencies', Dovey and Kennedy point out (ibid., p. 20). Technicity is also 'never fixed' due to a different reason; technicity is not a stable construct that we 'own' or fully decide, but rather something constructed in relation with others, similar to how gender is also a construction made in relation with others.

The concept of technicity will help to direct our focus at differences in access to construct, or to be recognized with, an identity that includes a close relationship to technology. As pointed out decades ago, by renowned (even canonical) researchers such as Turkle (1988) and Wajcman (1991, 2004), women have to give up part of their femininity when they enter the field of computing – they have part of their femininity 'undone'. This has also been illustrated in more recent studies, as in Adam et al.'s study of women in the UK IT workforce, where they report on women who downplay their femininity, or even 'forget' that they are a woman in the male-dominated IT business (2005, p. 288). However, as we will see in the following diffractive reading, it is not only gender identity or gender norms that can be 'undone', but also technicity, through processes of co-constructing gender and technology.

Thus, through a diffractive reading, focusing on the 'doing' and 'undoing' of both gender and technicity, normative as well as norm-breaking practices and articulations, I will look for patterns across the previous analysis, hoping that another landscape can be discovered, a landscape that can encourage more liberating research narratives (Haraway, 2004).

Short recap of the case studies

Before we continue with the diffractive reading, I will give a short recap of the most important lines and patterns found in the previous chapters, which will be the empirical framework for the ensuing discussion.

The discursive universe of a newspaper and a computer magazine explored in Chapter 2 illustrated how the personal computer entered Norwegian culture. It did not enter as a clearly masculine technology in the early 1980s, but rather with an ambivalent relationship to gender, where women appeared as potential users and IT workers in line with men. These stories disappeared during the 1980s, as a more masculine discourse grew stronger, gradually excluding women and covering up traces of women's relationships with computers. The masculine discourse continued to grow in strength until after the turn of the millennium, when the extensive spread of computers generated a growing number of stories documenting that 'everybody' uses the computer for 'everything'. The masculine discourse was clearly lurking in the background, thus, the inclusion of 'everybody' involved introducing characters directly in opposition to the young male nerd who used to dominate this space, and we find aunts, prostitutes and grandmothers as profiles of the new user groups.

Chapter 3 also has a historical perspective, but this time with the focus on traditional computer education, exploring discursive changes within recruitment initiatives since the early 1980s. The discursive changes are parallel to what we saw in Chapter 2, moving from a gender-blind discourse in the early 1980s, to a masculine discourse, to a feminized version of the masculine discourse and further, widening the discursive space by introducing new groups, in this case, social and non-nerdy men. This study illustrates that the understanding of gender-technology relations creates an important background for how to deal with challenges related to a low proportion of women in computer education.

Chapters 4 and 5 are based on observations and interviews among students of the Humanistic Informatics course, first in Chapter 4, exploring gender-technology relations among both male and female computer students, and again in Chapter 5, focusing on the women's expressions of pleasure about computing. This case study was carried out at the peak of the most masculine period of the discourse, and the hegemonic discourse emphasized men's relationship to computers. This case study documents variations between men and women, but it also illustrates variations within the group of women and within the group of men, as well as their opportunity to negotiate the hegemonic discourse in different ways. However, as long as the discourse benefits men more than women, negotiating the discourse is quite different projects for men and women. Chapter 5 gives an image of individual change and transformation among the female computer students, illustrating how most of the women who entered the computer course did so with certain (negative) expectations about their own relationship to computers. For most of these women, their negative expectations were not confirmed, and instead they developed their relationships with computers in positive ways, contradicting the discourse, entering a new landscape of a masculine and forbidden world.

The next section is devoted to exploring the findings across the contexts represented in Chapters 2 to 5. We have seen gender-technology relations develop, discursively as well as in practice, but how can we understand this; how and why do gender–technology relations develop as they do; and what effects does this cause?

Layers, adjustments and missed opportunities

The first notable pattern that becomes visible when reading Chapters 2 to 5 'through each other' is the different layers of meaning, seemingly partly contradictory, but still existing simultaneously in parallel. Rosi Braidotti has emphasized the importance of not only seeing differences between men and women, but also of recognizing differences between men and differences between women, and finally, differences within each individual (1994, p. 158). These layers of meaning coexist and might even result in what seem to be paradoxes. The first layer – differences between the genders – is obvious to most of us, and vital to how we constantly sort people into gender categories. This layer can often be too obvious in a way that makes it mask other differences, thus creating stereotypical narratives about men and women (Adam et al., 2004; Phipps, 2007). The second layer – differences within the genders – calls for a more nuanced understanding of gender-technology relations, requiring a letting go of the first layer of difference to allow for internal differences of the gender categories to appear. The third layer turns our focus towards awareness and recognition of paradoxes and what seems to be inharmonious connections or combinations of meaning in individuals' accounts, and to how individuals contextualize meaning in order to establish or activate different storylines. As Faulkner points out, most stories about women and technology are either positive or negative, failing to see women's ambivalence about technology (2001, p. 80), as if mixing these positions would introduce unwanted paradoxes. However, these different layers of meaning can help to explain such paradoxes, not as impossible, mutually exclusive positions, but as contextualized meaning. The different layers also produce important differences in how gender-technology relations are described or perceived, in particular in terms of how simplified and stereotyped or detailed and nuanced these relations appear. The newspaper case study in Chapter 2 illustrates how stories about gender-technology relations tend to be simplified, blurring variations beyond the hegemonic discourse, making the categories of 'men' and 'women' appear the most important, 'speaking' for men and women in general as two homogenous groups, while 'differences within' are less pronounced. On the other hand, the study of computer students (Chapters 4 and 5) includes several layers of meaning constructions. First, the students share the experience-based level of individuals, the 'me', which includes details, variations and even challenges to the hegemonic discourse. Second, the perspective of 'you' or the individual's environment, which also allows for more nuanced stories, and third, the discursive level of 'everybody' – the faceless masses most often described through stereotypes, simplifications and references to hegemonic discourses. The computer students articulate all these layers simultaneously. They relate to the hegemonic discourse, but in different ways, through different positioning strategies, constructing their own individual blends of technicity and gender identity, and in these stories they can be less faithful to the hegemonic discourse. The individual or local character of these stories means that the variations are not necessarily seen as likely to be valid as general descriptions. Thus, we find a computer course with a majority of women, but simultaneously it is questioned whether this is the 'real thing'. For general descriptions, the students rather refer to the faceless group of everybody and to stereotypes that they might not find to be true for everybody, but still likely to be true (enough) for the majority of 'everybody' – true enough to be used as general descriptions of large and homogeneous categories, such as men and women. The computer students mix these layers: some men and women 'use' the hegemonic discourse as it is (for men as a positive force and for women explaining why they fail), but most of them reconstruct their own local story in far more nuanced ways which might involve a challenge or even rejection of the hegemonic discourse. They illustrate how we 'do gender' for others, while we have the option of 'undoing' problematic aspects of gender-technology relations for ourselves or our nearest environment, however, in some cases only temporary or locally.

These three layers of the individual, their close environment and 'everybody' produce different kinds of meaning and different narratives about gender-technology relations depending on distance to the individual. The newspaper case study does not give the same admission to the close and nuanced perspective of individuals that we find among the computer students, which is the layer involving the most notable discursive resistance. The hegemonic discursive level acquires a more important position in the newspaper stories. When presenting potential nuances to this picture, 'discursively deviating' individuals, such as, for instance, female IT experts, they are presented in ways that restore their proper femaleness – for which stereotyping is a good tool. Stereotypes involve easily recognizable chains of equivalence and empty signifiers, working to homogenize and make a coherent story that we all can recognize (Laclau and Mouffe, 1985). When women engage in computing, they introduce a certain level of un-recognizability and incompatibility (cf. Wajcman, 2004, p. 112), or following Cockburn, when women engage in male work 'they upset a widely accepted sense of order and meaning' (1985, cited in Powell et al., 2004, p. 23).

Although the newspapers' tendency to simplify is well known, making the female IT experts more feminine has nothing to do with simplification, but rather with making them recognizable. In a study of women entering a male-dominated workplace, Edelman found that they were perceived as 'No ordinary girls' – unless they left the computerized job, in which case they were seen as 'just ordinary' women (1997). In this case, however, a balance is restored by emphasizing the female IT experts' feminine qualities, and they are constructed as 'ordinary' women, talking 'about weddings, children, men...about a lot of things that often occupy women' – information otherwise irrelevant to being an IT expert. This makes women able to enter computing and simultaneously remain 'ordinary women', without (seriously) threatening the discursive logic of women and computing as incompatible, and without threatening men's technicity (fascination, absorption, technical brilliance). The kind of technicity associated with women does not include the extreme fascination or absorption associated with men. In this act of creating discursive balance between non-traditional female technicity and traditional femininity, not only do the stories of the nuanced individual disappear through the act of making them seem recognizable, but also part of women's technicity disappears, leaving behind a modified and curtailed technicity, tied to being social and people-focused.

The perspective close to the individual also makes change more visible. The interviews with the computer students illustrate how they experienced change – first, on a practical level (in their own practice, skills, knowledge and so on) and, second, on a discursive level. The hegemonic discourse is challenged by the experiences they have as students, for instance, the presence of a majority of female students as well as female lecturers, or their own relationships to computers developing through discursively unexpected experiences, for women in particular through experiences of mastery and joy. Experiencing non-hegemonic situations challenges the discourse, but, for most of them only locally, temporary, for 'me' and perhaps for 'you', but not necessarily for 'everybody'.

Chapter 3, focusing on efforts to include and retain women in higher computer education is, as in the newspaper case, dominated by a shallow perspective uniforming through simplifications and stereotypes rather than revealing details and differences. The preconceptions in all the different discursive expressions discussed in Chapter 3 are firmly based on a stereotypical notion of gender – a simplified and dualistic image of gender, ignoring similarities between and differences within. These preconceptions make a direct impact on the way recruitment initiatives can be considered. Even attempts actively to shape the discourse in creative ways by emphasizing the importance of (the assumed feminine) social skills, leave the stereotypical notions of gender unscathed. The inclusion initiatives relying on this feminized discourse of computing have been severely criticized for their use of gender stereotypes (Lagesen, 2003; Lagesen and Sørensen, 2009), and one of the main problems with gender stereotypes is that they practically eradicate the possibility of dealing with individuals, with their individual backgrounds, motivations and stories. We might have to accept that it is difficult to plan for infinite variation among people, and that inclusion strategies for women somehow have to target women. Or at least, this has seemed like an unquestioned truth, but do inclusion strategies really have to target women as a group? In a study of gender policy in Norwegian schools, Gansmo found that school management who ignored the white paper's command to target girls with their computer inclusion policy and instead targeted all pupils could demonstrate a greater success in including everybody – also girls (2003). Indeed, several studies have pointed out that aiming for high quality for everybody, men included, is likely to be a better strategy than gender-specific solutions (Gansmo et al.,

2004; Faulkner and Lie, 2007), which, whether talking about computer education or computer games (Cassell and Jenkins, 1998, p. 36), are likely to re-introduce gender stereotypes and essentialism. By targeting the stereotyped group of women, the educational institutions are not able to deal with the perspective, or even the possibility, of individual, nuanced, partly traditional, partly non-traditional backgrounds, or of 'differences within'. It acts as a barrier for the close-up level that individuals use to adjust and construct their own positions – the positions that have the highest potential for dislocating the hegemonic discourse. The hegemonic discourse on its own can only make available the homogeneous categories of men and women, bogged down in sticky retrospective stereotypes.

We can see the tendency of the different layers of meaning producing different stories. The perspective close to the individual reveals numerous nuances and variations that include both acceptance, adjustments and rejection of the hegemonic masculine discourse. The shallow perspective, however, ignoring 'difference within', is incapable of reflecting nuances, and thus comes to rely on simplifications, stereotypes and other kinds of 'grand narratives' in which the categories of men and women remain homogenous. Consequently, the hegemonic discourse might seem less descriptive and normative in the nuanced perspective of the individual, and more like valid descriptions in the shallow perspective of 'everybody'. The nuanced perspective of individuals provides the potential to counteract the pressure of stereotypes and hegemonic descriptions that cannot act as truthful representations of individuals. These grand narratives are what Laclau refers to as 'myths' – as (necessary) simplified narratives that have the job of making the world perceivable and understandable (Laclau, 1990, cf. Torfing, 1999, p. 115). We cannot take all variations into account when referring to something without creating endlessly long and detailed stories. Thus, the job of 'myths', stereotypes, or hegemonic discourses is indeed to simplify, which in turn undermines the opportunity to re-tell stories that are faithful to real people. The shallow perspective is also more resistant to change, simply because it excludes nuances and variation. Not until 2003/04 do convincing stories about change in gender-technology relations appear in the newspaper, that is, not until the personal computer had reached a large majority of about three quarters of the population.[2] The nuanced perspective of the individual, however, is more inclined both to allow and to reveal change. Nevertheless, when individuals' stories are re-told within the remote and shallow perspective of the press, both nuances and changes tend to disappear again.

Braidotti's proposal to look for 'differences within' also means being aware of paradoxes, and instead of thinking about paradoxes as 'contradictions' that seem to exclude each other, paradoxes should be seen as a central part of living, of negotiating between hegemonic discourses and our own subjectivity, of 'what we do about what the world does to us' (Chapter 4, p. 107). Being aware of 'differences within', we might not only see that 'one size does not fit all' (Faulkner and Lie, 2007), but might also find that *one size does not necessarily fit anyone!*

Gender as difference

Gender, in the shape of dualistic differences between men and women, has become deeply embedded in the hegemonic discourse of computers that developed through the 1980s and 1990s, as illustrated in previous chapters. Descriptions of gender as difference not only dominate as the most frequently mentioned differentiating feature,[3] but also acquire standing as the most important origin of difference within this field, the one difference most difficult to avoid or from which to escape. What is described as the typical male relationship to computers is treated as a norm against which girls and women are measured (and found wanting) (Gansmo et al., 2003b), while the shades (or nuances) those that fall in between these categories have a tendency to remain invisible and be ignored. Men are allowed to represent an ungendered norm (Lohan, 2000), it has been claimed. Images of men and computer technology are, however, not similar to how images of men as ungendered representatives of 'man' have tended to appear in political or economical history (cf. Scott, 1988), where, almost as if by chance (or due to gender-blind research), representatives for 'mankind' are men. In relation to computing, men are not portrayed as ungendered, and it is no coincidence that they are men. Rather, the association between men and computer technology is closely connected to the fact that they are men, as has been illustrated by technology researchers across the Western world. We have in the previous chapters seen descriptions of technology and masculinity as defining each other (Wajcman, 2004), arguments that they need to be understood in relation to each other (Faulkner, 2001), and the claim that 'technology is what men do and women do not' (Oldenziel, 2001, p. 128). Thus, quite different from representing an ungendered norm, men represent a *gendered* norm in computing, and it is *as men* that they retain the discursively dominant position. It is boys' and men's close connection to computers that support the dualistic image of the hegemonic discourse. As computer users and experts, men are perceived as 'doing gender' in harmonious ways (Faulkner, 2005, 2006), while potential stories of men's negative or missing relationships to computers rarely reach the surface. A similar inability to see the whole picture is pointed out in *She Figures 2009*, as 'gender segregation in education is almost always presented from the perspective of the educational choices made by girls, even though gender segregation is also due to boys' preferences for certain fields of study' (European Commission, 2009, p. 67). When boys' preferences are treated as a norm, this cannot simultaneously be seen as a target for criticism in the same way that girls and women's preferences have to a certain degree.

Derrida explains how a dualism works as a system of exclusion, in which the poles of the dualism are defined as negations of each other, thus the poles are defined by being not-the-other. Feminists have theorized on this and suggested that we see women as 'the second sex' (de Beauvoir), 'the other' (Irigaray), 'zero', and 'not one' (Plant). Haavind sees gender as a system of two kinds of 'something' that is always split into masculine and feminine, and always with the masculine in a superior position (1993, p. 19). The content of these two 'somethings' can shift and gender can mean different things in different contexts, akin to 'empty shells' that can be filled with different things (Lie, 2003b, p. 277). This dualistic pattern that involves a split and a hierarchy also repeats itself on different levels, according to Haavind. Thus, while, for example, humanities and natural sciences are often seen as feminine and masculine fields, computer science can also be split into feminine and masculine parts, in a fashion similar to that scholars such as Turkle and Faulkner have illustrated, referring to hard/soft approaches to computer programming and engineering (Turkle, 1984; Faulkner, 2000a). A central feature of these dichotomies is, according to Faulkner, that they appear to be mutually exclusive, despite the fact that most engineers, male and female, have to deal with both people and technology, and they have to apply different styles of working. In short, both sides of these dichotomies are needed within engineering, meaning that there are significant mismatches between images and practices of engineering (Faulkner, 2000a, p. 85, 2001; Kleif and Faulkner, 2003).

When gender is perceived as a mutually exclusive dualism it contributes to a segmentation of gender as a difference in which men and women cannot be similar, and where gender (power) relations and inequality can never really change. When perceptions of men having a close, even intimate, relationship to computers are allowed to dominate the masculine pole of the dualism, the same dualism simultaneously segments the distance between women and computers, making women and computers appear as incompatible. One result of this is, as we saw in Chapter 4, women explaining their lack of computer skills by emphasizing that they are women: they are *not (computer-skilled) men*, or women laughing when being asked what they do to have fun with the computer, explaining that they are *not (computer-fascinated) boys*.

This dominating focus on gender as a dualistic difference prepares for and justifies gender-bound norms for men and women. It directs our view not only towards differences between men and women, but also towards *similarities* among women and similarities among men, making shades and nuances within the gender categories disappear. Non-hegemonic expressions become discursively invisible, unable to affect the discourse in more than highly temporary or local ways, as we saw in Chapter 2, where computer-using women as well as non-using men are ignored. While women (as a group of non-users) had to face the threat of having to go 'back to the kitchen sink', the rare non-using men – the 'Stone Age leaders' – simply 'risked' receiving their mail on paper. In Chapter 4 we met men who wanted to hide their lack of computer skills (that is, their lack of harmony with the hegemonic image of men's computer skills), which is both a consequence of the expectations of men in the hegemonic discourse, and contributes to stabilizing it by making male 'discursive deviation' harder to spot.

The discursive invisibility of women's computer use and skills as well as men's lack of use and skills stabilize a strong relationship between men and computer technology alongside a weak relationship between women and computers. Gender is continuously perceived as a mutually exclusive dualism, even when faced with norm-breaking evidences of men and women crossing the gender divide. One example of this is the researcher studying how boys and girls use the Internet, starting out with the claim that a previous gender gap in terms of access has gone but that there are, however, differences between how boys and girls use the Internet and concluding that 'it will always be like that' (Chapter 2). Changes revealed in the study hardly supported an everlasting gender difference, but still, that was part of the conclusion. Thus, the dualistic notion of gender can be activated even when faced with contradictory information, illustrating how gender tends to snap back into a familiar pattern, as well as how the dualistic gender perspective becomes a controlling force and acquires a discursively stabilizing effect.

If the starting point is that women and computers/computing are incompatible, one of them apparently has to change in order to alter the overall picture. During the 1980s and 1990s, both in the press and within computer education, the belief that women had to change was noticeable. In the press, women were asked to direct their interest towards computers. This is echoed in the governmental report quoted in Chapter 1, which urges women 'to take more interest in technology and industry'. This request is based on a notion of technology as neutral; the problem lies with women, not the technology (Cockburn, 1999 [1983], p. 127). A related strategy, expressed within the masculine discourse of computing (Chapter 3), starts from the notion of men and women as different, although, assuming women have the ability to learn to become more like boys (the norm) when relating to (the neutral) technology. Thus, equality measures aimed at helping women to acquire 'boy's room competence' appeared to be a logical solution. We have to stop asking what is wrong with women, and start asking what is wrong with computing, Håpnes suggested in the early 1990s (1992, p. 156). The most creative strategy we found within computer education, the strategy aimed at enlarging the ethos of computing by adding 'female qualities' in order to invite women, can be seen as an attempt to change computing and not women. This change, which was more rhetorical than practical, resulted in, among other things, dissatisfied female students (Lagesen, 2003), having been invited to a study that did not live up to the slogan: 'The computer science programme is more about human beings, than about machines'.[4]

The gender dualism's effect on technicity

In the slogan, computing is assumed to be something for women that is different than it is for men; it is a way to practice social skills. On the one hand, the technological aspects disappear, but on the other, women are assumed to be able to bridge two normally distinct fields: social skills and technological skills. However, in the process, women's technicity is constructed as socially-based, different from men's technologically-based technicity. Men understand technology – women understand people, and through the work of the dualism, the opposite side is excluded (Lagesen, 2003; Faulkner, 2005). Thus, even though women (can) participate in technological contexts, they do not make any notable imprint on technology, because their technicity is already defined as socially-based, not technically-based. Women's weak ability to leave a 'mark' is certainly not a general feature of women, as many other professions or activities have gradually become associated with women or femininity as more women have entered the field, such as pre-school teaching, nursing or even technological activities such as blogging. Despite being computer users, women are not necessarily recognized as 'proper' computer users. During the early 1980s secretaries were clearly recognized as the office's leading-lights on computers, but their computer use did not affect the hegemonic discourse – illustrating how, instead of gender being undone (no one has claimed that secretaries had to give up part of their femininity), it was rather the female secretaries' technicity that was undone. Related undoings have been reported for other female pioneer groups in computer history, such as the women programming ENIAC. '[T]he job of programmer, perceived in recent years as masculine work, originated as feminized clerical labor', Light claims (1999). As we saw in Chapter 2, the ENIAC pioneers did not think of themselves as pioneers, but simply 'doing the job we had been hired to do' (Gürer, 2002, p. 119). Neither the ENIAC women nor others experienced their job as pioneer programmers as technical, and according to Nathan Ensmenger it was not until the 1960s that the community of programmers was 'actively making itself masculine' (2010, p. 116), and eventually, programming turned into an activity with clear masculine associations. Women's abilities and skills being ignored or made invisible is recognized in several other cases (Edelman, 1997; Lie, 1998; Adam et al., 2005; Staunæs and Søndergaard, 2008). An example can be found in Skelton's study of women in academia who experienced that their gender overshadowed their professional status: 'they see a young woman and immediately assume I'm the secretary' (Skelton, 2005, p. 325). In all these cases it is not women's gender identity being undone, but rather their technical or professional identity.

The recruitment campaign at NTNU made an attempt to tie femininity to ICTs by activating a discourse that already included femininity: about communication and caring. This campaign is not unique in doing so, and by the early 1990s, Sørensen had already asked whether an increased proportion of women would bring increased influence of feminine caring values into technology research (Sørensen, 1991, 1992). Sørensen indicates that 'female engineering students are somewhat more care-oriented than male engineering students'. However, he concluded that 'women in technological R&D do not (any more than men) support caring through work-related norms and preferences', due to 'educational and workplace socialization' that makes 'female R&D scientists, relatively speaking, less inclined to show caring values' (Sørensen, 1992, pp. 15–16, 19). Although female values are wanted in technological professions, the effect of including women is not totally clear (Mack, 2001). Female traits can also create a double-bind for women, as Staunæs and Søndergaard illustrate in their study of gender and top management (2008). Diversity has become a key word to management in the private sector in Nordic countries, the authors claim, tied to a hope that bringing in 'diversified groups of employees…will improve' the organisation in vital ways (Staunæs and Søndergaard, 2008, p. 135). A similar perspective on diversity is also found in other Western countries, and a continuous failure to attract women to IT work is presented in the US as representing a series of 'risks' (Ashcraft and Blithe, 2010; see Chapter 3). In practice it is, however, not that straightforward, and it has also been illustrated how feminine traits are appreciated and valued as mastery if shown by men (Woodfield, 2002; Staunæs and Søndergaard, 2008). However, the paradox appears when women display femininity, as it is not perceived as mastery, but rather as femininity 'out of control'. This creates a catch-22 for women, in which they have to choose between being 'interpreted as uncontrolled femininity-carriers, something that will question their talents as managers', or 'renounce their privileged access to femininity' which leaves them with 'nothing particular, nothing "different", to bring in', which activates the question: 'Why hire women who act like men, when you can hire men?' (Staunæs and Søndergaard, 2008, p. 153). Stuanæs and Søndergaard elegantly illustrate how difficult boundary-making can be when assumed feminine qualities become a virtue when displayed by men, but a problem when displayed by women. A similar dilemma can be related to computing; why hire a woman who is good with people if what you need is someone good with technology? Woodfield illustrates how this paradox appears in practice in the IT business, arguing that women are invited into the computer business based on their feminine values, but when work is to be done, it is still the technical and 'masculine' values that win (Woodfield, 2000, 2002).

Many of the female computer students interviewed (Chapter 4 and 5) illustrated the paradox with which some of these women live, when they claim that they do not feel inferior to boys or men (cf. Powell et al., 2006), simultaneously as they meet a world that perceives them as inferior or less technically capable than boys and men (cf. Adam et al., 2005). Is it possible to protest against this undoing? The newspaper study did not provide any examples of secretaries protesting against the undoing of their technicity, but we met a young girl, angrily objecting to all girls being called 'retards' in relation to computers: 'Don't call us retards one more time!!!!!!!!!!!!!!!!!!' (Chapter 2). The newspaper article claiming that 'girls are retards' in relation to computers clearly violated the image she had of herself and her friends, all of them active computer users, and the 18 exclamation marks signal a need to 'shout' her protest. Among the female computer students (Chapter 4) we saw a similar need to 'nag' to be heard and an active claim to the right to be perceived as women that are fully-competent with computers. While some women clearly protest against the undoing of women's technicity, most of them do not. When the lie 'girls are lousy with computers' is 'repeated time after time, you believe it', one of the women claimed, clearly aware that it was a lie, but also aware that it affected her (cf. Hayes, 2010b).

After the millennium, when the hegemonic discourse fell out of step with observable practices, the discourse was challenged in more general terms in the newspaper, with boundaries re-established to allow for a more diverse variety of user groups. It is, however, clear that in this process of re-drawing the discursive boundaries and letting something new in, something else had to go: 'transformation cannot affect only the pole of "the others". It must equally dislocate the position and the prerogative of "the Same", the former centre', Braidotti claims (2002, p. 14). The first that had to go in this case was the hegemonic image of nerdy boys and young men. The boys and men who used to be the computer wizards and leaders of the future, those who most clearly were barriers to women's entrance, stand out as the main target to fight in this process of re-defining the discourse. The 'weapon' used against the image of the nerdy men is that which previously was definitely beyond the discourse and different female groups previously quite (discursively) unlikely to be proper computer users, such as your aunt, elderly women and prostitutes put on the barricade to fight the image of the young nerdy men. The way these young male geniuses' hegemonic position as computer wizards is undone in this process is a rare example of men's technicity being attacked. Computer wizardry is, however, no longer tied to knowledge and skills acquired in the boy's room, but to 'recipes' for how to do harmful and criminal hacks, available online for anyone to use.

In Chapter 2 it was claimed that the PC entered Norwegian culture at an intersection between different fields, relevant in different contexts and to different groups. While this diversity could have fostered an 'interpretative flexibility' (Pinch and Bijker, 1997 [1987]) providing a diversified image of computer use and users from the start, what actually happened was that the intersection rhetoric activated stabilizing mechanisms by putting everything into a black box. By avoiding drawing boundaries between different fields and levels of computer use, certain contexts, aspects and users/experts were allowed to become authoritative representations in a black and white picture with flattened nuances and variations. The discursive changes since the millennium with the introduction of 'everybody' provide a refreshing destabilization of the intersection rhetoric, by introducing nuances and varieties that do not fit in the black box any longer. And the new boundaries drawn after the millennium also indicate that women are allotted a new sense of technicity.

Redoing femininity, undoing technicity?

'Feminine identity', Wajcman claims, 'has involved being ill-suited to technological pursuits. Entering technical domains has therefore required women to sacrifice major aspects of the gender identity' (Wajcman, 1991, p. 164; 2004, p. 112). What we have seen here, however, suggests that it is not always gender identity that women have to sacrifice, or that this sacrifice is something 'required' from women, but rather that it is women's *technicity* that is contested. In many ways gender seems to remain rather stable (Lagesen, 2003), while access to technicity is disputed, and gender acts as a barrier to women's technological activities being acknowledged as technicity. The claim that women have to relinquish part of their femininity when entering computing has acted almost as a truth in its own right, but perhaps it is time to re-evaluate this, and instead focus on how women's technicity has become a contested area. For instance, how would recruitment strategies look if we take this knowledge into account? The incompatibility hypothesis supports the idea of recruitment strategies aimed at supporting women's feminine side, as we have seen in one of the recruitment campaigns. What if we focused more on women's technicity instead of femininity? The 'boy's room competence course' presented in Chapter 3 partly does that by inviting women to increase their technical skills, but through a highly exclusive metaphor of the 'boy's room' which involves an interpretation of girls and women's computer skills as inferior. Is it possible to imagine a recruitment campaign focusing on women's extreme fascination for technology or their brilliant technological skills? (I write this hoping that in a decade or so, someone will find this sentence hilarious and outdated, just as I hope Cynthia Cockburn appreciates that her vision from 1983 is outdated today: 'How impossible it seems to imagine a technical training course... where women are simply in a majority' (1999 [1983]). If you still think this is impossible – read Chapter 4 again).

Underestimation of women's computer knowledge is one of the significant barriers women encounter. It has been found that women also tend to underestimate their own computer knowledge (Sølvberg, 2002; Cheryan et al., 2009). They start to 'believe the lie', and even to distrust other women's knowledge, or see other women as a potential threat to their own hard-earned 'inside-position' if they were to put on a show of an expected female relationship to computers (Nordli, 2003; Powell et al., 2004, 2006; Godfroy-Genin and Pinault, 2006). We know that achievements are closely tied to expectations, and when someone is clearly and openly expected to achieve poorly, that is what they do (Steele, 1997). Women are expected to achieve poorly in relation to computers, and in Chapter 4 we found this tendency reflected by several of the women, among them, one who actively used the stereotyped expectations to explain her lack of understanding, and another woman using the same stereotypes in a more positive way, claiming that she could easily impress her family with her computer skills because so little was expected from women.

Men, however, are associated with computers and computer skills, and we have seen men claiming that computing is something they learn more easily because that is what is expected from them. The expectations of men even make it possible to appear as computer competent under false pretences, such as one of the male students who apparently acquired this status primarily due to being a man, but without really having the proper competence. He did, however, realize this mismatch himself, thus claiming to feel safer in the female-dominated environment at Humanistic Informatics than he would have felt in a male-dominated group of computer students. Thus, both men and women seem to have an awareness of how stereotyped expectations affect their relationship with computers, as well as how they are perceived by others. Non-skilled male users have certainly not received the focus they deserve from research. However, on an individual level it remains a more severe problem that women lose their technicity, thus in some cases also their professional identity, than men being attributed skills that they do not have.

What is missing from this picture? None of the female computer students' stories, none of the newspaper stories and certainly none of the recruitment initiatives within computer education gave any examples of women being perceived as having more computer knowledge than they had. Being over-rated as computer users is not a readily available option for women (cf. Søndergaard, 1996, p. 156).

How can gender appear as stable?

One of the most repeated claims of our time is that, even though we might not understand it, and the direction might be unclear, one thing that is clear is that everything is changing (Braidotti, 2002; Gere, 2008). But still, gender keeps coming back as a stabilizing force. How can gender appear as stable when everything changes? One reason already discussed is the fluidity of gender, which means that the content can change simultaneously with the notion that gender as an everlasting differentiating feature remains strong. Another and concomitant reason is that people live as men and women, and most of us need to be men and women in ways that make us recognizable. Already in childhood girls and boys learn that they need to do gender (West and Zimmerman), and they learn to perform gender (Butler) in certain ways, through clothes, friends, their laughter and not the least, in how they relate to technology. Most people also want to do gender in clear and unambiguous ways; we want to be recognizable in order to avoid exclusive or non-inclusive reactions (cf. Søndergaard, 1996, p. 33). That does not mean that doing gender is the same thing always, or everywhere, or for everyone. We can do gender in non-traditional ways, but in order to remain recognizable, a balance between traditional and non-traditional gender expressions is necessary, according to Søndergaard. The new freedom of the individual in the post-traditional society (Giddens, 1991) and the fluidity of 'liquid modernity' (Bauman, 2000) do not free us from gender. 'Sexed identity...is tattooed on bodies', according to Braidotti (2002, p. 26), and we cannot choose not to express gender (Søndergaard, 1996, p. 35). For all the interviewed computer students, to avoid being recognized as men or women was not only impossible, but also not desirable. For the women it was important to be recognizable as women, and many of them employed a distinct not-boy rhetoric. In one sentence they could describe how they enjoyed using computers, while in the next sentence they rejected the idea that they could have fun with computers: only boys could do that. Even the most discursively progressive women emphasized their not-boy qualities. However, neither did they apply the kind of feminization strategies we saw in the press. Being compared to a girl was not seen as positive to either men or women. Girl, in this setting, includes associations with 'girly' and stereotypical images of girls, images that otherwise might be resented or discouraged, just as Cassell and Jenkins *warn* us against 'teaching girls to act like girls are supposed to act' (1998, p. 28).

Doing gender is not difficult. We all do it every day, and as Butler illustrates, becoming gender is about repeating recognizable patterns, about 'transferability of the attribute' (2004, p. 213). The repressing aspect of this is that we tend to repeat what is recognizable, which for women, among other things, has involved a distance from technology

(Rommes, 2007). The liberating side is that we can, in fact, introduce new meaning. More women present in computing education makes the choice less strange for a woman, the female students explained. The catch is that women doing technicity is not easily discursively recognized as such (Mahoney, 2001). Being a woman has been found to be a barrier to being perceived as computer competent, and 'successful SET professionals are not perceived as feminine' (Bagilhole et al., 2008). The discursive changes at present can be compared to a lever where a majority has to be on the 'right side' before it tilts and for women to become recognized as computer users. This is quite different from male technicity, where barely 12 per cent of men using the Internet made men as a group qualify as 'Internet users' in 1997 (Chapter 2).

'The time is ripe for reworking the relationship between technology and gender', Wajcman had already claimed in 1991, pointing to how '[t]he old masculinist ideology has been made increasingly untenable by the dramatic changes in technology, by the challenge of feminism and by the new awareness of the vulnerability of the natural world' (Wajcman, 1991, p. 166). Wajcman emphasizes feminist politics as a driving force for change, while she criticizes Sadie Plant for, in a vein of technological determinism, putting too much weight on technological development as the driving force in improving women's situation (Wajcman, 2004, p. 76). 'It becomes clear', Plant writes, 'that if the ideologies and discourses of modern feminism were necessary to the changes in women's fortunes which creep over the end of the millennium, they were certainly never sufficient to the processes which now find man, in his own words, "adjusting to irrelevance" and becoming "the disposable sex"' (Plant, 2000, p. 274). According to Plant, technology achieves what feminism could not. Some of the important agents of change we have seen here and need to add to this picture are people themselves. New technology does nothing until people use it, and it is only when people change their behaviour that the world will change (Shirky, 2008). But we have also seen that changing practices alone are not enough. They also need to be recognized discursively, and for that to happen, it seems as though it is the masses, not the exceptions, that count. Another important agent of change, in particular in Western countries, is the laws of gender equality. The strong governmental involvement in creating gender equalities in the Nordic countries, in particular Norway and Sweden, has been referred to as 'state feminism' (Hernes, 1987). Laws and regulations are strong agents of change that can affect material conditions very directly. Discourses, however, are more delicate to work with. Trying to change peoples' thoughts and perceptions is more often associated with censorship and less democratic regimes. However, some of the East European countries show a higher proportion of women in science, technology and engineering (Charles and Bradley, 2006) because 'the former Soviet ideology officially promoted gender equality but no gender awareness', according to Godfroy-Genin (2009, p. 85), while the freedom of women in Western countries has also been translated into a 'right to choose poorly paid female-labeled career paths' (Charles and Bradley, 2006, p. 196).

In recent years, it has been emphasized that there is a need to re-introduce matter, materiality and the body into feminist theory (Braidotti, 2002; Barad, 2007; Smelik and Lykke, 2008; Grosz, 2010). 'Methodologically, the return of "real body" in its thick materiality spells the end of the linguistic turn in the sense of the postmodernist over-emphasis on textuality, representation, interpretation and the power of the signifier', Braidotti claims (2006, p. 50). A focus on discourses does, however, not imply an insistence that they exist independent from materiality. Discourses are constructed, strengthened, questioned and rejected by real people living real lives, and neither does materiality exist independently of discourses; our construction and perception of society go through discourses. We do not build a computer, design programs for it or use it independently of discourses that provide meaningfulness. It is through discourses that we understand and perceive the material world, our bodies, identities, doings and undoings, and these same things produce and yield meaningful input to discourses. Discourses and materiality are mutually dependent, intertwined and they affect each other. Both can act as stabilizing forces and both can act as agents of change, which we can clearly see in the case of the personal computer. The arrival of the personal computer was followed by a strong impression of the computer being widespread much earlier than it really was, thus, with its potential presence being able to push the discourse. The actual diffusion of computers and computer skills among women during the 1990s and early 2000s, however, remained discursively hidden, and did not push the discourse before a large majority had acquired and started to use computers in the years after 2000. Suddenly we all had mothers, sisters, daughters and aunts using computers. Even though we perhaps could expect that materiality, with its solidity, would outplay discourses, one of the main stabilizing forces seems to be related to categorization and stereotypes, exclusion and inclusion tied up in a heteronormative pattern that forces gender back into a binary. Alternatives remain invisible, or when they are highlighted, they need explanations or modifications. One of the major challenges is that 'reality' does not beat the discourse. Reports about women and girls using computers or computer games as much, or more, than boys and men often produce reactions in online forums such as 'bullshit', 'I don't believe it', references to alternative interpretations, alternative conclusions, or critique against the research methods.[5] 'Scratch the surface of any discussion list about female gamers', Dovey and Kennedy write, 'and you will soon uncover a common discourse that condemns these players as outsiders who have "misappropriated" the game space and who potentially undermine gameplaying expertise' (2007).

Different layers, different contexts, competing discourses as well as materiality create a complex and layered image of change and stability. Discourse and materiality might be different entities, but it is the pattern they make together that makes our living environment.

Stability and change, hand in hand

It is important to ask what the stabilizing forces are, but it is just as important to ask what has changed and what might be destabilizing forces, 'to trace the moments where the binary system of gender is disputed and challenged', inharmonious 'breaking points' and moments of discontinuities (Butler, 2004, pp. 215–16). There is a 'complex interplay between what replicates the same process and what transforms it', Foucault points out (quoted in ibid., p. 216), which we have seen several examples of here.

What patterns do we find pointing towards change and stability? First, we have to remember that a discourse has the ability to appear as fixed and stable at the time, disguising change through the discursive logics that makes it appear coherent. Thus, we find popular discourses, as coagulated stereotypes, conservative, blind to variation and ignoring women's technicity. But we also find popular discourses reflecting change, which becomes visible from a long-term perspective. Through the 1980s, 1990s and after 2000 there are notable changes in the hegemonic discourse of computer technology, moving from a gender-ambivalent notion, to a masculine, and further by taking up the battle with the hegemonic masculine discourse. Focusing on individuals, we find stabilizing forces making women and men perceive themselves and others in light of the hegemonic discourse, but we also find considerable variations between individuals, and change within individuals. Female computer students can describe themselves as *atypical* women because they have chosen computer education, simultaneously as they describe themselves as *typical* women within computer education. The first refers to themselves compared to other women (traditional women who choose typical women's education), while the second refers to themselves versus men within computer education. While studies of popular discourses might require a long-term perspective to reveal changes, the study of real people requires that we get close enough to see the 'difference within' – within the gender categories and within individuals. Within educational institutions we see how attempts to create progressive and innovative images of computer education activate forces of change (invitation to women) as well as stabilizing forces (traditional gender stereotypes). Change and stability go hand in hand, also in images of female IT experts, challenging the hegemonic discourse by their presence, and serving to stabilize gender-technology relations by emphasizing their femininity and difference from male computer experts. However, one of the most pronounced stereotypes, and also the one in which most hope has previously been placed – the emphasis on women's social skills as the reason for inviting them into computing – primarily exists on the discursive level. When women finally arrive, feminine skills are not necessarily celebrated (Woodfield, 2000) in women (Staunæs and Søndergaard, 2008). Instead, women are taught to become less like women, as the mentor project presented in Chapter 2 illustrates, where a mentor described his student's success as having become 'less emotional'. It has also been pointed out that the presence of 'more women as designers or programmers does not necessarily result in products better geared towards women ... nor that female IT professionals are more communicative' (Maass et al., 2007, p. 14).

'Neither gender nor ICT nor their relationship may be assumed to be constant', Sørensen reminds us (2002b, p. 9). However, one of the challenges seems to be the way gender remains stable, based on traditional images of femininity. The gender-technology entanglements explored here require a more complex understanding than a dualistic image can provide. What is clear, however, is that the hegemonic masculine discourse cannot provide faithful images of men and women's relationships to computers. I have already indicated that 'one size' might not fit anybody, and perhaps there are other categories that could explain individuals in a more faithful way than gender? And perhaps in that way we can be computer users first, and gender second?

7
Is There an Elsewhere?

Introduction

In Chapter 1, I questioned the way the 'stability argument' concerning women's relations to technology has re-emerged within feminist technology research during the last decade. Claims that 'nothing has changed' for women in computing, that earlier improvements have stalled or eroded, and that things are 'not likely to improve any time soon' (Cohoon and Aspray 2006a p. 139; see also Chapter 1, p. 2) appear parallel with claims emphasizing social as well as technological change. In this chapter I will ask what stories we tell about gender-technology relations, and how we tell these stories. What is allowed in and what is left out? Are there other ways of telling them? I will repeat the stability argument: certain things have not changed. But I will also modify it, as it involves one specific way of reproducing the narrative about how gender-technology relations have developed.

We have seen examples of policymakers being criticized for ignoring diversity within the gender categories when dealing with gender and ICTs, and we have seen a tendency to ignore diversity within ICTs and computer-related education. Sally Wyatt has suggested that we should allow technological determinism to re-enter into technology research, in recognizing that technological determinism is often included as a vital part of the story told by actors we study (2008b). And perhaps we need the same for gender: most people still refer to gender as a dichotomy, thus we also need to be aware of how gender categories are used as a dichotomy, simultaneously as we are faithful to the diversity within, the doing as well as the undoing of gendered categories, norms and subjectivities. Would it still be possible to find 'pure difference' or an 'elsewhere' where gender-technology relations could develop in positive ways?

Numbers matter, but what matters to numbers?

Technology became masculine when a definition of engineering as white, male and middle-class took hold during the late-nineteenth century, Wajcman claims, and in the process 'diminishing the significance of both artefacts and forms of knowledge associated with women' (2004, p. 16). The modern concept of technology appeared, and at 'the same time, femininity was being reinterpreted as incompatible with technological pursuits' (ibid.). Optimists of the 1970s and early 1980s, however, suggested that information technology was different from earlier mechanical technology, and that women would not be excluded from the new, clean and silent technology that could not be reserved for men based on arguments about bodily strength (Wajcman, 1991, p. 150). As we have seen here, in the early period of the personal computer the gendering was ambivalent; it was presented as a new technology bringing new opportunities for both men and women (Chapter 2), and it could symbolize different things and support different gender identities (Lie, 1998). However, the consolidation of a masculine-connoted discourse during the 1980s and 1990s also consolidated computer technology as 'gender inauthentic' to women (Faulkner, 2000b), and women 'as an additional element in a technological education, rather than an inherent part' (Salminen-Karlsson, 1999, p. 219). Thus, the optimists were partly right, in that women also could be included in the information technology revolution. However, the discursive detour of the 1980s and 1990s weakened the link between women and computers, and women appeared to be excluded. Exclusion has been a central topic within research on women and computer technology; women's exclusion from technological contexts, education, work, leisure time, and as we have seen here, exclusion from the hegemonic discourse. Exclusion also lies at the heart of the 'stability argument' presented in Chapter 1, where, among others, Wyatt was quoted, claiming that 'gender inequalities persist, even in countries such as Norway' (2008a). Wyatt is right; even in Norway, the number of women choosing computer science is still low, but as I will argue here, we need to ask what women are excluded from, and the nature of the exclusion mechanisms put to work here.

One of the most important exclusion mechanisms, most frequently blamed for the persistent low number of women, is male dominance of computer technology. This involves domination in several ways, through history, in production and development as well as use, as passive domination and as active resistance to women's entry in technology-related contexts (Cockburn, 1985; Wajcman, 2004). Sørensen, in categorizing gender and ICTs research, describes this as a 'storyline' defining women's exclusion as 'a story of a strong, long-term cultural momentum that causes women to be kept out of technology, ICT included. Men does [*sic*] not want women in technology, they have struggled for a long time to keep them out, and this will be difficult to change' (Sørensen, 2002b, p. 26). This storyline is supported by Cockburn's claim from the early 1980s, that '[w]omen are not merely failing to enter technology. On the one hand we are being repelled, and on the other we are refusing' (1999 [1983], p. 127). In some ways it all seems to be connected to men's domination in numbers, and because women continue to be a minority in science and engineering, 'the outcome is a culture of technology where women and femininity appear as matter out of place' (Sørensen, 2002b, p. 22).

Already young girls begin to 'sense' this male dominance, Nissen has proposed, making them want to stay away (1996). Women also choose to stay away, Cockburn claimed in the early 1980s: 'We are not failing, we are on strike' (1999 [1983], p. 127). A decade later the situation had not improved, and Webster claims that the hostility towards women in the IT workplace culture results in women 'electing not to enter it' (1996, p. 177). 'The door may be open, but the world beyond it does not invite entry', Mahoney writes from the US (2001, p. 171), and from the UK, Adam et al. claim that 'it is remarkable that any women at all persevere with an IT career' (2005, p. 283). How can we understand this long-term exclusion? Are girls and women still 'on strike'? Do girls still sense the male dominance? And do women still meet active resistance to their entry?

The male domination of computing has not only originated from men's dominance in numbers, but has also been reinforced as a result of how women's participation has, to a considerable degree, remained hidden and invisible (Misa, 2010a), not counting, and therefore unable to affect or correct stories of male dominance. There should be no doubt that making women's contributions to technology visible is important – a task feminist researchers have been working on for several decades (Abbate, 2003), and which has given tools and data to rewrite the his-story into a story that also includes and is moved by women (Gürer, 2002). Second, girls do not want to stay away from computer technology any longer. 'We tend to think about technology in terms of industrial machinery and cars, ignoring other technologies that affect most aspects of everyday life', including women's close encounters with technology, Wajcman pointed out in 1991 (p. 137). At the turn of the millennium it is, however, no longer obvious that 'industrial machinery' dominates our image of technology. Instead, one of the dominating images of technology is tied to digital technology. The diffusion of computer technology, including the Internet, has reached the same level as radio and TV in Norway, with 92 per cent coverage for computers and 91 per cent for the Internet in 2009 (Vaage, 2010). Although Norway is in the lead regarding the diffusion of ICTs, other Western countries have also experienced a diffusion of ICTs to a majority of their population, with a spread of Internet access to 82.5 per cent of the UK population and 77.3 per cent in the US in 2010[1] (cf. Chapter 2).

For girls and boys there is practically no difference in terms of access to computers, in Norway as well as in other Western countries (Becta, 2008). ICTs, however, are no longer homogeneous things. 'The generic concept of ICT is less meaningful to young people', Gansmo et al. write, and the 'issue is no longer whether or not to use ICT, but what activities you need ICT to do' (2003b, p. 137). Computers used to belong to 'the realm of machinery and mathematics – a daunting combination for girls', according to Wajcman (1991, p. 152). Since the 1990s ICTs have developed in a number of other realms, having shifted 'from being an exotic technology' to 'become a mundane part of our everyday practices and everyday technologies' (Elovaara, 2004, p. 13). Technological changes have also produced the foundation for new relationships to ICTs, with increased usability for computers and computer applications, and mobile phones that can no longer simply be used for making phone calls (Jenkins, 2006, p. 5). New mobile phones, palmtops and e-readers make it possible to connect to the Internet wherever the user is, with access to blogs, Twitter, Facebook, YouTube and a number of other sites and services for socializing, networking or sharing. There is a long and constantly growing list of technological gadgets and activities that do fit neither the image of technology as 'industrial machines', nor that of mathematics. It is no longer working-class men that dominate our images of technology, but rather the youth, the generation that has grown up with ICTs, who take them for granted and have the ultra-quick typing of an SMS in their thumbs. The younger generations born since 1980 have been called 'Digital Natives', different from the older generation of 'Digital Immigrants' (Prensky, 2001). The Digital Natives 'all have access to networked digital technologies' and 'they all have skills to use those technologies' (Palfrey and Gasser, 2008, p. 1). To describe the new generation's relationship to new media the word 'use' is no longer sufficient, but instead we should talk about 'produsage', Axel Bruns suggests, where 'use' and 'production' merge (2008), in a convergent culture where new patterns of collaboration and creativity create the basis for new kinds of networks that affect old patterns of power (Jenkins, 2006; Shirky, 2008). In a survey of gender stereotyping of ICTs skills among the 'Millennials' (those born between 1982 and 2000), Trauth et al. found signs suggesting 'that the tight grip that masculinity has held on the IT field might be loosening as ICT becomes increasingly ubiquitous and embedded in the everyday life of both males and females' (Trauth et al., 2010, p. 7).

The foundation for individual relationships to technology has changed, and with it, we see changes in gender-technology relations. The male-dominated history of computers and girls' 'sense' of a male heritage does not make them stay away today, if it ever did. Girls and women have moved heavily into the world of social media, such as the blogosphere, Facebook and other services that are used not due to need or instrumental choices, but because people want to use them. Women's participation in new social media clearly indicates their active engagement as well as their pleasure in computer activities.[2] Unfortunately, these are also fields where women's technicity is easily withdrawn, leaving only the non-technical side of the activities, such as writing a diary (similar to the old-fashioned paper diary), communicating or being social.

Optimists have predicted that if girls were presented to computers they would become interested (Faulkner, 2001, p. 79; Gansmo, 2003, p. 139), which according to the dominant narrative did not come true; increased access to and use of computer technology has not, to date, raised girls' interest in IT education. ICTs presented at school are seen as 'less interesting and important' by both girls and boys, and simultaneously 'it seems as if gender fades away', according to Gansmo et al. (2003b, p. 133). This might indicate a closing gender gap in relation to ICTs. However, confining the fun and motivating use of ICTs to leisure time activities indicates that this might be 'a more important learning arena' than schools, in particular one in which more boys than girls develop an interest for computer science education (Gansmo et al., 2003b, p. 133). Thus, what is perceived as boring use of computers at school might even reinforce girls' perception of computing as boring, also reinforcing the tendency of computing being male-dominated, fed by boys who discovered their fascination and motivation for computers in their leisure time activities. In the case of the computer students we have met here, we have seen that some of the women started with a perception of computers as boring, as one of the women illustrated with the equation 'computers = technical = boy things = boring'. However, most of them also experienced increased pleasure in computing with increased knowledge, in particular in knowledge about 'the real stuff'. Programming and working with technical aspects of the computer – tasks defined as masculine by the hegemonic discourse – were among the activities about which the women expressed the greatest pleasure. They not only enjoyed doing it, but also how they had acquired skills to do something they had thought was not meant for them, something that put them on a level with men. They enjoyed the fact that they could understand what men were talking about, that they could impress their friends, and that they acquired a bit of the status they associated with these tasks. They dislocated a particular aspect of the gender-technology relation: the association between lack of computer skills and women, or the weak link between women and technicity. But they also kept the gender dichotomy alive. It was the unexpected connections between women and computer skills that made them able to surprise and impress. Some of the women openly expressed their love, addiction and fascination for the computer, but not all. Some of them, such as the women who laughed out loud at my 'ridiculous' suggestion that they could have fun with computers, were making rhetorical restrictions to their pleasure in computers, similar to those that have been found in previous research (Håpnes and Rasmussen, 2003). Thus, we still find examples of women modifying the descriptions of their use of and pleasure in ICTs, but we also find an increasing openness for women to express deep pleasure in technology.

The most common solution applied in an effort to fix the low proportion of women in computing has been attempts to increase the number of women (Webster, 1996, p. 177). However, within the most technical subjects of computer science and informatics this has not yet been achieved. Closely related to this solution to the gender divide is the assumption that only if we reach a 'critical mass' (Kanter, 1993 [1977]) of women in computing will the problem disappear.[3] First, it will mean that women can relate to other women instead of being a minority among men, thus giving both social and professional support to women. In Chapter 3 we saw the special 'girls groups' designed to give women 'like-minded' computer students to discuss matters with. In Chapters 4 and 5 we saw that female students were both surprised and relieved upon finding a female majority in Humanistic Informatics; surprised because they had expected women to be a minority, and relieved because it made it less strange to be a woman within computer education. This indicates that a larger number of women can make computer education appear as less 'gender inauthentic' for women. It also indicates that the existence of a critical mass is important, not only for social pleasure or professional mastering, but in relation to women's image of themselves as recognizable. Reaching a critical mass might create new discursive boundaries. Powell et al., however, criticize the idea of a critical mass as a way to improve women's situation in engineering, claiming that 'women engineers either share the values and attitudes of their male colleagues or that on embarking on an engineering career, women assimilate to the engineering culture, failing to challenge the dominant masculine discourse' (Powell et al., 2006, p. 696). Reaching a 'too high' proportion of women is not necessarily good either. Salminen-Karlsson claims that a previously male-dominated field tends to lose its high status if too many women become involved (1999).

Initiatives to recruit and retain women in computer science in the late 1990s had temporary success, but the numbers dropped again after 2000. Thus, the critical mass argument is difficult to sustain because of the slow or stagnating growth, which is different from the position of women in a number of other previously male-dominated fields that have developed steadily into having an equal share of men and women, or even into being female-dominated (Hayes, 2010b). It also indicates that increasing the number of women, although it did change the picture at the time, was not enough to feed a continuous movement towards equal proportions of men and women. This effect could be observed at NTNU in Norway, where some major initiatives and campaigns to recruit women during the late 1990s did succeed in increasing the number of women. However, at NTNU the increase of women did not rely on a previous critical mass of female students, and the critical mass of female students temporary established through the initiatives did not create lasting changes, that is, a lasting critical mass.

It is also necessary to ask for whom a critical mass will work. While some women clearly appreciated finding a majority of female students in Humanistic Informatics, some of them did not see it, and some did not see it as particularly positive. One of them pointed out that she had not even taken particular notice of the majority of female students, but instead noticed the majority of male lecturers, and another woman said with regret that the female majority did not really count, because Humanistic Informatics was not like the 'real thing', such as informatics or computer science. The size of a 'critical mass' can be disputed, and will clearly be of varying size in different situations and contexts. It is, however, not the size of a critical mass that matters, according to

Powell et al., but rather what the proportion of women represents, as for instance scattered and isolated women would make 'a false number': 'The precise number is, therefore, less important than the nature of the response that the new minority receives from the majority' (Powell et al., 2006, p. 692). If we consider how the diffusion of computers had to reach a three-quarters majority before female user-groups became clearly visible in the popular discourse of the press (see Chapter 2), this might perhaps indicate that the critical mass of women related to computing is higher than previously thought. If a critical mass is what is needed for women not to feel uncomfortable due to being a minority, what we really need in order to change the image of computing is a mass with the ability to mark and leave discursive traces. Lecia J. Barker and William Aspray refer to a study showing that 'children's perceptions of whether occupations were male or female' needed a mass of 70 per cent of either men or women holding those positions for the occupation to be considered male or female (2006, pp. 35–36) – a proportion similar to that needed before female computer users became discursively visible in Norwegian popular discourse.

Even though women make up more than 60 per cent of all students in Norway today, the proportion of women within the categories of technical and computer-related educational programmes has not shown any particular increase since the turn of the millennium.[4] So, yes; I, too, will have to repeat the stability argument: the number of women is still low in these educational fields, despite all efforts to increase the proportion of women. The most frequently used stability argument, 'the number of women in computing remains low, despite ...', clearly points to a problematic field. But we should also be careful not to let the stability argument become a disturbing, if not destructive, force in the research discourse about gender-technology relations. The stability argument tends to focus on one single symptom in one single field, which has also been simplified, so it contributes to the exclusion, masking and eradication of other perspectives and understandings. Even though we can agree that numbers show stability (cf. Charles and Bradley, 2006), and that 'numbers matter' (Lagesen, 2007), we also need to ask: 'What matters to numbers?' What do the numbers really show? Is it enough to talk about computing as computer science and informatics? As we have seen, the intersection rhetoric collapses the different layers of contexts, uses and users related to the personal computer and allows a minority group (computer-fascinated boys and young men) to define a norm (Chapter 2). A similar thing happens when we talk about computer education, which has developed from informatics and computer science into a number of different educational branches, many of them within social sciences and humanities, such as Humanistic Informatics, where we can find a higher proportion of women. Within the official national statistics of Norwegian students' choices of education (NSD), Humanistic Informatics is not counted as a computing subject, but rather as humanistic education in the same category as languages, art history and philosophy. As long as the right to be counted as computer education is reserved to studies under faculties of natural sciences and technology, computer studies at other faculties that attract more women remain invisible in this story. Thus, we partly fail to see women's participation in computer education when ignoring the contexts where they participate, denying them entry to the discourse of computing. This does, however, not release us from continuing to problematize the low proportion of women within certain areas of computer-related education, but it does mean that we could *talk* about women and computer education in another way. We *could* have told a story of computer education where women participate, thrive and become computer experts, where they could find 'like-minded' and female role models, and where '[i]ntense pleasure in skill, machine skill, ceases to be a sin' (Haraway, 1991a, p. 180).

Another aspect often forgotten that could also contribute to another story is related to what Peter Denning and Andrew McGettrick call the myth about computer science, reminding us that it is not only women who do not choose to study computer science. Computer science also has problems recruiting boys, in Europe as well as the US (British Computer Society (BCS) e-skills UK and Intellect, 2009). Denning and McGettrick suggest that this is due to a myth that equates computer science with programming (2006), which makes it appear as less attractive not only to girls, but also to many boys. Large numbers of boys and young men choose not to study computer science, but this group is rarely met with gender-specific explanations or recruitment initiatives such as those targeting women, and the male disinterest in computing is to a large degree (discursively) invisible.

Symbolism, stereotypes and perceptions are clearly still problematic for women in technology (Faulkner, 2001, p. 79). They represent what is perhaps the most problematic field because it seems to work as a barrier to acknowledging women's actual activities with technology. Negative stories are what most effectively undo women's technicity, and one of the main barriers for women in relation to computing seems to be negative expectations of women's relations to technology (Barker and Aspray, 2006, p. 32). Girls and women still meet disbelief in the combination of femininity and technological skills, and, as we have seen here, they might even 'believe the lie' that girls are bad with computers. Even when they know it is a lie, they still have to face it. One way to undo negative symbols, stereotypes and perceptions is to fuel the dominating narratives with alternative stories (Hayes, 2010b). We often hear that women lack role models in computing (Kekelis et al., 2005; Cohoon and Aspray, 2006a; Ashcraft and Blithe, 2010). However, women can only become positive role models if they are not perceived as an 'out-of-role'-model, absorbed by a masculine culture, as having changed sides, as I was by one of the female computer students. Neither can women act as good role models if they are constantly being questioned, as the legendary female hacker Netochka Nezvanova whose true gender identity was questioned,[5] or the way reports about an increasing number of women playing computer games are met with disbelief, rejection or alternative explanations. There needs to be something for a role model to resonate with in order for the role model to act as such. We might also have to question the kind of role models produced when female IT experts are re-feminized at the cost of their technicity. Denise Riley points out that collective identities, for example 'women', are impermanent and alternating, and they are not the only identities we have. It is more common, she suggests, to 'skate across the several identities, which will take your weight, relying on the most useful for your purpose of the moment; like Hanif Kureishi's suave character in the film *My Beautiful Laundrette*, who says impatiently, "I'm a professional businessman, not a professional Pakistani"' (Riley, 1996, p. 31). For computing to become less gender-inauthentic to women, we not only need female role models, but female role models who are allowed to be professional computer experts.

Can there be pure difference?

Gender-technology relations are not stable (Sørensen, 2002b, p. 9), but rather constantly in motion, developed discursively as well as materially, and 'the relationship between technological and social change is fundamentally indeterminate', Wajcman points out (1991, p. 163). Discourses are shaped in the meeting between us and our environment, through our articulations, actions and perceptions. The meaning is, however, not uniform or flat. We deal with meaning in a number of layers and contextual situations, whereof some count for the individual herself, some for her close environment and others for 'everybody'. The last 25 years of science, technology and society (STS) research has challenged notions of technological determinism that sees technology as developing on its own technological terms (Wyatt, 2008b), and concepts of co-construction have dominated the last decade of STS research. Discourses and technology are intertwined, affecting each other. However, we have also seen that they might be out of tune with each other, and that discursive change can be made by the mere indication of a new technology on its way, as in the early 1980s when the discursive diffusion of personal computers outstripped diffusion in numbers. We have also seen that a massive material change might need to go before a discursive change, such as the discourse not embracing female users before diffusion of the computer had reached a majority of women.

Braidotti paints a picture of our contemporary society as schizophrenic. In its simultaneous celebration and rejection of change, 'it combines the euphoric celebration of both human and technological gadgets, *new* wars and *new* weapons with the complete social rejection of change and transformation' (Braidotti, 2006, p. 2). She finds a similar paradox in how difference is increasingly encouraged – '[p]ost-industrial societies make differences proliferate to ensure maximum profit' – simultaneously as traditional mechanisms of grouping and homogenizing people continue. As a result the 'unity of the subject of humanism is exploded into a web of diverse discourses and practices. This phenomenon, however, seems to leave miraculously unscathed the centuries-old forms of sexism, racism and anthropocentric arrogance that have marked our culture' (ibid., p. 44). Braidotti warns against believing that we can remove gender as a difference. The problem is rather that difference has been turned into something negative. 'Like a historical process of sedimentation, or a progressive cumulation of toxins, the concept of difference has been poisoned and has become the equivalent of inferiority: to be different from means to be worth less than. How can difference be cleansed of this negative charge? Is the positivity of difference, sometimes called "pure difference", thinkable?' (ibid., p. 4). The question for Braidotti is not one of difference or not, but whether there can be a positive difference, a difference that is not evaluated in hierarchies, and where opposites can exist without one becoming the norm and the other becoming devalued. Thus, she insists on keeping the notion of gender as a difference, and to see complexities and differences also within the nomadic subject, which 'is marked by a structural non-adherence to rules, roles and models', opening for 'disengagement and disidentification from the socio-symbolic institution of femininity' (ibid., p. 40). This reflects the findings here rather well: neither men nor women we have met in the previous chapters reject their own gender identity – they might adjust, change, strengthen or moderate it, but they do not reject it. Thus, Braidotti's question is important: can we create room for 'pure difference' – a difference that is not entangled in hierarchies of de/value and social status?

We might also ask what the real importance of gender difference is, as Connell reminds us that studies of differences between men and women most often find a larger percentage of similarities than differences (2002), but still it is the differences that attract our interest. There is a large number of studies of 'computer use, attitudes and experience', according to Barker and Aspray, however, many of 'these studies are designed to look for differences and describe probabilities, not similarities nor generalities' (2006, p. 29). Similarly, in a review of research literature on gender and information systems, Adam et al. found that, in particular, quantitative research has a tendency to reproduce dichotomous and essentializing pictures of gender (2004). The problem of being stuck with large categories such as men and women is that there are no ways of talking about them other than as differences between men and women. These dominating ways of talking about gender also count for the people we study: they also talk in terms of gender differences, but they are also able to juggle different layers of meaning, to talk about the large and homogeneous categories simultaneously as they adjust their individual and local stories in which gender takes on a more varied skin, and where hegemonic gender norms are being challenged. The individual and local stories do not, however, as easily reach or affect the hegemonic discourse, but it is our responsibility as researchers to listen to these variations, variations that can challenge the hegemonic discourse.

Research as storytelling

Scientific representations or stories are both performative and material, according to Haraway (cf. Asdal et al., 1998, p. 174). They produce what they report about, such as 'female', 'a highly complex category constructed in contested sexual scientific discourses and other social practices' (Haraway, 1991a, p. 155). The 'narrative practice' of science is not built on innocent revelations of truth, but rather involves mechanisms in which 'fact and fiction coshape each other, to be a reduction of their hard-won science to subjective storytelling' (Haraway, 2008, p. 312, n29). Scientific discourses give shape and meaning to reality, and Haraway encourages us to retell the world in liberating ways. Which stories do we make when analysing gender-technology relations, and how can we retell these stories in liberating ways?

I do not intend to devalue questions or worries about the minority position women have in certain computing fields. I do, however, want to emphasize the importance of seeing what women actually do. By following women to where they are, we will find women being educated and working as computer experts. In the new ITEC professions women professionals work in a wide range of the newest areas of computer and technology-related occupations, with e-logistics, e-content, software and digital art, among others (Webster, 2006). There are 'many different entry points for the new ITEC professions', Webster points out, and 'not all of these are conventional, formal educational channels' (ibid., p. 6). The danger of the 'stability argument' is that, once again, it makes many of these women, their contributions to technological development and their technicity invisible. Many things have changed during the last 30 years – the period so often referred to as one of stability for women in computing (Webster, 1996; Corneliussen, 2003b; Cohoon and Aspray, 2006a; Wasburn and Miller, 2006; Wyatt, 2008a). Among the general changes in the academic world are the increased presence of women and a less clearly informal patriarchal academic culture that makes it less likely that a woman with lipstick becomes 'hotlips' and a library for women's studies the 'menstruation room', all reported by senior female professors[6] (Niskanen and Florin, 2010). Among the more formal changes are regulations of women's participation in committees and boards, rules, laws and strategic plans dealing with gender equality and career development for women in academia. And finally, if we open up the definition of computing to other fields than informatics and computer science, we will find studies with a higher percentage of women (cf. Chapter 3), documenting women's wish to work with technology. We should not fall into the pit described by Skjeie and Teigen (2003), where we appreciate movements towards the goal while accepting continuous inequality, but neither should we forget about the changes, even the small ones. The rhetoric of stability related to women's position in computing seems plausible because there is still an extensive job to do, but at the same time this rhetoric locks in changes and improvements, and continues to construct women as 'not really interested' in technology (cf. Chapter 3, Margolis and Fisher, 2002).

We can also hope to instigate change in the hegemonic dualistic stories of gender-technology relations through research optics. Different research perspectives can produce different stories, and as we have seen, a perspective close to the individual will make visible difference and change that often remain hidden in and by the hegemonic discourse. Thus, the nuanced perspective focuses on individuals has a potential for producing alternative stories to the oft-repeated story of women's incompatibility with or disinterest in technology, and we might find women in love with technology and women becoming computer experts and enjoying it!

Gender as a differentiating dualism is one of the most conservative and preserving forces in the discourse of computers. One of the main challenges seems to be to create a new non-dualistic discourse, or to follow the 'utopian dream of the hope for a monstrous world without gender' (1991a, p. 181), as Haraway formulated it. This is perhaps also the most difficult challenge, as most people seem to find the essentialistic and dualistic image of gender true to how they perceive the world. During the spring of 2009, a TV series in Norway created a fierce debate about gender research and the question of what makes men and women different. From this debate we can see that many people find arguments building on biology and essentializing claims to be more trustworthy and more in line with how they see the world, quite different from social scientists' notions of gender as socially constructed. Readers' comments in newspaper debates following this TV series warned against 'tampering with nature', illustrating a lack of trust in social sciences along with a strong belief that natural scientists are more likely to produce true research results because they are not producing them as such, instead they are simply unveiling facts (Barad, 2007, pp. 40–1). This line of thought is conservative and discursively stabilizing because attempts at changing things appear to be attempts to change nature (Scott, 1988, p. 60).

Although this TV series also created debate among gender researchers, most feminist researchers today would (still) renounce essentialistic explanations to gender differences as being rooted in and fixed by biology. However, many perceptions of gender differences related to ICTs established in the dominating discourses appear as 'truths' that apparently do not need to be confirmed through research. They have acquired an impression of nature, and even within gender research a default application of gender as a dualism might appear (cf. Chapter 2). When the 'default gender' is applied, we might miss the opportunity to rewrite the story, even when reality provides material useful to change the image of women as non-technical.

Elsewhere within here

Gender identity, femininity and masculinity – it is something we do or perform, we dress in it, talk like it, walk like it – make it materialize.

But it is also a feeling, a perception, an idea, norms, discourses. The same goes for technicity. We do it, think it, feel it and make discourses in relation to it. One important, but often ignored field is feelings. We have seen that women have developed stronger relationships to computers since the 1980s, not only in terms of having become comfortable computer users, but also in an increased discursive openness to *feel* comfortable – to enjoy computers. Sherry Turkle's female interviewees in the 1980s felt confused, irritated or even angry with how they felt about computers (Turkle, 1988), almost as if they disliked the fact that they liked computers. Research in Norway has also documented girls' reluctance to admit that they enjoy computers (Håpnes and Rasmussen, 2003). 'Only boys can have fun with computers', some of the female students told me, while simultaneously talking about their own quite extensive and pleasurable use of computers. Many of the female computer students we have met here did, however, experience a profound change in their relationship to computers, where, apart from increased computer skills, perhaps the most important change was their new feelings: they experienced love, fascination, addiction and excitement. Thus, women do act in new ways today, but more importantly, they also feel in new ways. They move boundaries, change their universe and widen their playground to include new recognizable images of being women – women who have positive feelings for technology.

Against 'worlds at war' , Haraway has 'tried to write into a more vivid reality a kin group of feminist figures', hoping they 'might guide us to a more livable place' called 'elsewhere' (Haraway, 2004, p. 1). This is 'a place composed from interference patterns' (Haraway, 1992, p. 300), and a place for 'inappropriate/d others', an optics suggested by Trinh T. Minh-ha (1989) for exploring difference and otherness, referring to 'not quite other, not quite the same.'[7] 'Elsewhere within here' is the here and now, Trinh T. Minh-ha explains in an interview: 'while one is entirely involved with the now-and-here, one is also elsewhere, exceeding one's limits even as one works intimately with them'.[8] We know today that women, despite the stories of their technological incompatibility, can love technology for itself, that they can be eager computer gamers (Nordli, 2003) or enjoy programming (Abbate, 2010), and that they can be the majority in a technical class. We also know that women, as much as men, comprise a diverse group with a diverse set of preferences, likes and dislikes. We need messy identities, whether cyborgs or monsters (Haraway, 2004), that can reject the pure identities that so neatly fit into the gender dichotomy. We need messy identities to support our diversity and multiplicity, and to avoid our images of gender and technology ending up as copies of copies without an original (Baudrillard, 1994).

The starting point for this book was the repeated claim that the situation for women in computing had not changed substantially since the 1980s. We have seen that this is in part true. Many things have changed, but there is yet more to achieve in terms of equality in gender-technology relations, both discursively and materially – even in Norway. But how special is Norway in this picture? We have seen a number of 'surprising' similarities between Norway and other Western countries, in the low proportion of women in computing as well as in discursive formations generally negative to women. The similarities are surprising due to high expectations of gender-equality in Norway (cf. Chapter 1), which suggests that we could also expect 'more' when it comes to gender-equality in computing in Norway. We have also seen Eastern European countries scoring lower on gender-equality measures than many Western countries, but still documenting higher proportions of women in computing (Charles and Bradley, 2006). Thus, Norway is an interesting case for the study of gender equality in computing, in particular because of the high expectations of general gender-equality, resulting in statements such as 'even in Norway' when pointing out the stability regarding women's situation in computing.

At the same time as we have seen many similarities among Western countries, local contexts are also important. A working solution in one context might not work in another due to local or national culture, Trauth et al. point out, claiming that gender mainstreaming politics possible in Europe do not hold an equally strong position in the political culture in the US (Trauth et al., 2009, p. 493). However, as we have seen here, politics alone does not seem able to change discourses or stereotypes that have been recognized as barriers to women. It is also important to notice that inclusion strategies are not the same as exclusion strategies reversed, as pointed out by Faulkner and Lie (2007). Thus, the road to gender-equality in computing is probably not as linear as proposed by Ahuja, to simply identify barriers to women, and '[o]nce identified, these barriers can be addressed and appropriate solutions to overcome them can be found' (2002, p. 20).

As pointed out before, any effort to create changes in gender-technology relations needs to recognize that 'men and women fall into more than two categories' (Maass et al., 2007, p. 16) and that 'one size does not fit all' (Faulkner and Lie, 2007). To achieve gender-equality in computing, not only in Norway, but internationally, we need the 'elsewhere' – the more 'livable place' (Haraway, 2004, p. 1) to become more visible, to allow for the 'differences within' to become appreciable, and where 'pure difference' can exist. Women already inhabit the elsewhere within here: they already 'do' their technicity, but they also still experience the need to explain, adjust and justify their technology relations in order to be seen and respected. Women cultivate their own technicity, even when it is not recognized as such, and even when they have to nag or shout to be heard. The main challenge at the turn of the millennium is not to make girls and women become interested in computers, but to ensure that their relations to technology are not constantly being undone by images of femininity on the one hand, and of images of boys' and men's relations to technology on the other. We need a world where women do not have to defend or explain their position in computing, and where women can be professional computer experts without being forced simultaneously to take on the role as professional women.

Notes

1 Disrupting the Impression of Stability

1. *Aftenposten*, 4 June 2004.
2. *Meta-Analysis of Gender and Science Research* is a project of the 7th RTD Framework Programme of the European Union (http://www.genderand-science.org/).
3. Although Norway scores high on participation in working life for both men and women compared with most European countries, Norwegian women more often have part-time jobs than the European mean, according to Statistics Norway (SSB). In the early 1990s, 48 per cent of women and 9 per cent of men had part-time jobs, while the numbers for 2006 were 42 per cent and 12 per cent, respectively (Forskning.no (April 2008) http://publisering.forskning.no/artikler/2008/april/180052, date accessed May 2009, SSB Labour force statistics (April 2009), http://www.ssb.no/english/subjects/06/01/aku_en/, date accessed May 2009).
4. The Global Gender Gap Index 'benchmarks national gender gaps on economic, political, education- and health-based criteria'. The intention of the Index is to measure 'gender-based gaps in access to resources and opportunities in individual countries rather than the actual levels of the available resources and opportunities' (Hausmann et al., 2010).
5. This project involves the 27 EU member states, as well as Norway, Iceland, Israel, Switzerland, Turkey and Croatia.
6. Numbers of 'gender occupational segregation' refer to 'the share of the employed population that would need to change occupation (sector) in order to bring about an even distribution of men and women among occupations or sectors' (Meulders et al., 2010, p. 21).
7. A good starting point for obtaining an overview of early feminist technology research is Judy Wajcman's *Feminism Confronts Technology* (1991), which could also be read together with her more recent book *TechnoFeminism* (2004), where she looks back and continues the project she started in 1991.
8. In Norway there was, for instance, a regulation of pregnant women's use of computers at work, due to how computer work was assumed to stress the foetus.
9. There is a chronological error in this argument which Plant has borrowed from Gerald Edelman (*Bright Air, Brilliant Fire*, New York, Basic Books, 1992), as the Turing machine was described in 1936, almost two decades before the hormonal treatment (Leavitt, 2006).
10. The domestication theory was subsequently developed (Berker et al., 2006), and has also received a great deal of interest in Norway (Sørensen, 1994; Aune, 1996; Sørensen, 2006).

11. According to Ormrod, men were pictured as both single and family men, while women were only pictured as family women (1994).
12. The concept of the cyborg was first used by Manfred Clynes and Nathan Kline in 1960 to describe the relationship between humans and technology as a presupposition for people to survive during space travel (Haraway, 1997, p. 51).
13. Men occupy approximately 80 per cent of positions as leaders; women earn approximately 80 per cent of what men earn; more women than men work in part-time jobs, women dominate within higher education, although men dominate in certain technological subjects, and so on. (cf. Statistics Norway, http://www.ssb.no/english/; Skjeie and Teigen, 2003).
14. See Abbate (2003), for an extensive list of this research.

2 Changing Images of Computers and its Users since 1980

1. Women's invisibility is partly due to deliberate choices to make them invisible, as described by Thomas J. Misa: 'Two of ENIAC's women programmers were removed from the scene – literally cropped out of the picture – when the U.S. Army placed the ENIAC machine with one male operator at the center of its 1946 recruiting campaign' (2010a, p. 252).
2. Augusta Ada Byron translated Luigi Menabrea's report on Charles Babbage's Analytical Engine in the 1840s. She added her own notes, four times as long as the original report, in which she included the first sketches for a computer program (Campbell-Kelly and Aspray, 2004). Although most of what Lovelace wrote was Babbage's thoughts, according to Martin Campbell-Kelly and William Aspray, they were presented in her voice. Others, however, have pointed to Lovelace as the first to describe the computer as a creative and aesthetic machine, as she compares its use to the Jacquard loom: 'We may say most aptly, that the Analytical Engine weaves algebraical patterns just as the Jacquard-loom weaves flowers and leaves', she writes in her notes to Menabrea's report, http://www.fourmilab.ch/babbage/sketch.html (accessed January 2010).
3. Grace Murray Hopper, Admiral in the US Navy, developed the first compiler, and she was vital in the early development of programming languages (Gürer, 2002; Misa, 2010a).
4. One of the frequently repeated anecdotes from this era is when Steve Wozniak, bound by contract to offer his employer, Hewlett Packard, any new electronic inventions he conceived, presented them with his new and crude version of the first Apple computer, whereupon they, according to the movie *Pirates of Silicon Valley* (1999), rejected his invention with the stupefying question 'Why would anyone want a computer in their home?'
5. *Aftenposten* is Oslo-based, but read by people in all regions of the country, and is a mainstream newspaper of record that is widely respected. The main empirical material is gathered from *Aftenposten*, and focuses on more than 200 news articles that in some way discussed or examined computer technology *and* gender. Also, the Norwegian computer magazine *Datatid* has been explored. *Datatid* was published between 1979 and 2000 with 10–12 issues each year.

6. Some claim that the first personal computer was Altair, a computer-building kit, made available in 1975 (Ceruzzi, 1999, p. 72). 'Personal' in this case meant that it was cheap enough for 'anyone' to buy, and it could be operated by one person alone. However, calling Altair the first personal computer also emphasizes the importance of technical interest and curiosity, rather than practical need or usefulness, as those who managed to put the pieces of Altair together in the correct order would get a box that 'could do little more than ... blink a pattern of lights on the front panel' (Ceruzzi, 1999, p. 75).

7. The most famous computers were the Apple I, Apple II, the Apple Macintosh and IBM's PC. IBM's PC was launched in the US in 1981, but it did not reach the Norwegian market until 1983 (Nerheim and Nordvik, 1986). However, the IBM PC was important to the development of personal computers in several ways. It created a norm for technical development and a number of 'IBM clones' followed in its wake (Ceruzzi, 1998, pp. 277–278), and 'PC' became one of the most widely-used names for a whole generation of computers. The Norwegian translation 'Personlig Datamaskin', commonly used at that time, should make the abbreviation 'PD', which was also recommended by the Norwegian Language Council (who even encouraged the press to withhold parts of the salary for writers using English computer terms (*NTB* 7 July 1988)) but that did not catch on, and PC is still a common name for the computer in Norway. Counting the occurrences of 'PC' in *Aftenposten* (in the digital database *Atekst*) over the period analysed shows seven hits in 1983, 94 in 1990 and 858 in 2000.

8. *Datatid* 1979 no. 1.

9. 9 July 1995. Unless another source is named, all dates given in this format refer to *Aftenposten*. All quotes from Norwegian newspapers and magazines are translated by the author.

10. 14 January 1983, 11 November 1983.

11. 31 January 1984.

12. 13 December 1982.

13. 7 November 1981.

14. *Datatid* 1979 no. 5, 1981 no. 2, 1982 no. 8, 1985 no. 3, 1987 no. 11, 1991 no. 12.

15. *Aftenposten* 7 November 1981.

16. 8 May 1985.

17. 12 January 1996, 5 October 1997, 7 March 1999.

18. 22 February 1985, 24 December 1985, 11 July 1986, 17 March 1987.

19. 17 October 1996, 11 March 1999.

20. 30 March 1994.

21. 29 August 1997.

22. 5 August 1982, 11 March 1983, 5 February 1986, 24 September 1999, 3 October 1999.

23. 10 February 1987.

24. 7 January 1996, 29 May 1998.

25. 23 October 1985, 5 March 1987, 22 May 1989, 5 February 1997, 5 October 1997.

26. 16 December 1987, 16 October 1994, 5 September 1997, 7 March 1999.

27. 29 August 1997, 11 March 1999.

28. 5 October 1997.

29. Ibid.
30. 8 June 1997.
31. 5 March 1988.
32. 16 October 1994. This focus is also highly evident within research on men's relationship to computers (for example, Haddon, 1991; Nissen, 1993; Mellström, 1999; Turkle, 1984; Kleif and Faulkner, 2003).
33. 26 November 1988.
34. 15 December 1996.
35. Ibid.
36. Ibid.
37. 15 February 1986, 26 November 1988.
38. 28 July 1994.
39. 2 May 1995.
40. 13 December 1982.
41. 2 May 1995.
42. 26 November 1988, 12 November 1996, 27 January 2000.
43. 30 June 1991, 25 June 1994, 1 November 1994.
44. 23 November 1995.
45. 29 February 1992, 29 September 1993.
46. 24 May 1985, 11 April 1987, 26 January 2000.
47. 5 February 1992, 25 June 1994, 16 March 1995, 27 January 2000, 18 May 2000.
48. 15 December 1996.
49. 4 October 1998.
50. 12 June 1993.
51. 25 September 1993.
52. 26 March 1995.
53. 5 October 1997.
54. Ibid.
55. 17 December 1985, 16 October 1994, 4 August 1996.
56. 16 October 1994.
57. 4 August 1996.
58. 14 August 1996.
59. 2 December 1985.
60. 8 September 1989, 14 August 1996. This was not, however, supported by research from the late 1980s (cf. Rasmussen, 1988).
61. 17 December 1985, 15 May 1986, 17 March 1987.
62. 26 March 1995.
63. 22 November 1999.
64. 11 April 2000.
65. See Charles and Bradley (2006) for a cross-national comparison of the gender composition of computer science programmes in Norway and 20 other industrial countries.
66. *Computerworld* 14 August 2008.
67. 10 February 1987.
68. 24 December 1985.
69. 11 July 1986.
70. 16 December 1987, 16 October 1994, 4 August 1996.
71. 9 May 1997.

72. 5 October 1997.
73. Ibid.
74. 16 March 1997.
75. A similar change in media portrayals of teenage boys as computer experts has been pointed out by Sherry Turkle in the twentieth anniversary edition of *The Second Self* (2004, p. 215).
76. 25 January 2001, 16 November 2001. There are no examples of girls being associated with computer criminality, and there is no equivalent phrase to 'boyish pranks' that refers to girls.
77. 7 October 2001. There are also some articles about paedophilic activity from this period, referring to pictures or videos 'saved on a PC'. However, the computer is not involved in discussions of gender-technology relations in these articles, thus they have not been included here.
78. 4 June 2009.
79. 25 March 2001, 11 April 2001, 11 August 2001, 23 September 2003, 28 March 2006, 18 March 2007, 09 July 2007, 02 October 2007.
80. Office for National Statistics, 2010, UK http://www.statistics.gov.uk/ (accessed December 2010).
81. *Internet World Stats* http://www.Internetworldstats.com/ (accessed December 2010).
82. 11 March 2005.
83. 28 March 2006.
84. 11 April 2001.
85. 14 March 2005.
86. 4 June 2004.
87. Ibid.
88. 22 March 2004.
89. 23 July 2006.
90. 23 July 2006.
91. 12 September 2004.
92. 2 November 2003.
93. Access to the Internet varies across the globe, and according to the *Internet World Stats* statistics for 2010, the percentage of the population with access to the Internet is 58.4 per cent for Europe, 77.4 per cent for North America, 21.5 per cent for Asia and 10.9 per cent for Africa. Divided by countries, Internet access for 2010 is 51.7 per cent in Italy, 62.6 per cent in Spain, 68.9 per cent in France, 82.5 per cent in the UK, 97.6 per cent in Iceland and 77.3 per cent in the US (http://www.Internetworldstats.com/, accessed December 2010).
94. 29 April 1999.
95. *Dagens Næringsliv* 18 May 1999.
96. 29 April 1999.

3   Discursive Developments within Computer Education

1. The overall proportion of female students in Norway was 60.9 per cent in 2009, according to NSD (http://dbh.nsd.uib.no/statistikk/, accessed December 2010). In 2010, the worry was rather that women 'are taking over the universities', raising the question of whether 'women will eventually

dominate the academic disciplines' (http://eng.kifinfo.no/nyhet/vis. html?tid=71207, accessed July 2010).

2. 'Census Bureau Reports Nearly 6 in 10 Advanced Degree Holders Age 25–29 Are Women', US Census Bureau (http://www.census.gov/newsroom/releases/ archives/education/cb10–55.html, accessed 06 December 2010).

3. *Computerworld* 24 October 2003.

4. See, for instance, http://www.acm.org/announcements/pipeline.html: 'The new "Incredible Shrinking Pipeline Unlikely to Reverse" study from ACM-W has found that from 1983 to 1996, the percentage of women earning bachelor degrees in CS shrank from a high of 37.1 percent to a low of 27.5 percent' (accessed 13 July 2010; Camp, 1997).

5. *Aftenposten* 4 October 1998.

6. *Computerworld* 14 February 2003, and personal email correspondence with project manager Erik Arisholm, Simula Research Laboratory.

7. See Lagesen (2003) for an in-depth analysis of the first three advertising campaigns from 1997, 1998 and 2002.

8. The numbers below serve as a sample to illustrate women's participation in computer and technology related subjects at Norwegian Universities. Due to changing educational structures after 2000, as well as rapid development particularly in technology related subjects, it is difficult to trace precise numbers of students at different study programmes. All numbers are from Norwegian Social Science Data Services (http://www.nsd.uib.no/nsd/english/index.html, date accessed January 2011).

   Percentage of female registered students:

   - IT – Language, Logic, Psychology (Faculty of Humanities, University of Oslo): 2004 – 33; 2005 – 36; 2006 – 30.
   - Department of Media and Communication (Faculty of Humanities, UiO): 2003 – 60; 2010 – 61.
   - Centre for Technology, Innovation and Culture (Faculty of Social Sciences, UiO): 2003 – 61; 2007 – 62; 2009 – 52; 2010 – 42.
   - Department of Informatics (Faculty of Mathematics and Natural Sciences, UiO): 2003 – 18.5; 2007 – 18; 2010 – 17.
   - Bachelor in New Media (Faculty of Social Sciences, University of Bergen): 2008 – 51; 2009 – 47; 2010 – 48.
   - Bachelor in Information Science (Faculty of Social Sciences, UiB): 2008 – 21; 2010 – 19.
   - Department of Informatics (Faculty of Mathematics and Natural Sciences, UiB): 2003 – 14; 2006 – 7; 2010 – 13.

9. As with Björkman (2005), Mörtberg (2000), Bratteteig and Verne (Bratteteig, 2002; Bratteteig and Verne, 1997) in the Nordic countries, or Grundy (1996) and Lander and Adam (1997) in the UK, and Crutzen (2009) in Netherlands, and the list could be extended.

4 Variations in Gender–ICT Relations

1. Humanistic Informatics grew considerably in the late 1990s, but did not offer a separate study programme between 2003 and 2010, thus, the number

of students fell. A new bachelor's degree programme under the new title of Digital Culture was opened in 2010, and by autumn was attended by 11 male and 15 female students.

2. Humanistic Informatics has its roots in the early 1980s and a time when it appeared important to know how to program a computer, even for researchers and teachers in the humanities. Thus, programming had an important place within the curriculum, continuing until well after 2000. Humanistic Informatics changed its name to Digital Culture in 2009, and programming is no longer considered core knowledge in the study programme, although practical knowledge and skills are still important elements of the courses, thus many classes involve student activity, including programming, in computer labs.

3. My double role as both lecturer and researcher (which is discussed in depth in Corneliussen, 2002) made me careful not to discuss the research project in class. The good atmosphere and the extensive time I spent in the lab with the students gave me confidence that this double role was not problematic for the project (or the students), but rather a resource. Being a lecturer gave me a natural position among the students, thus, I could observe the participants in their everyday context of being computer students while trying not to 'disturb' this context with my position as a researcher (Lie, 1998, pp. 43–4).

4. At the time of this study, it was still not unusual to have students with very little prior experience of computer use in general. Thus, a 'crash course' was offered at the start of the first term, to teach students the most basic computer skills, such as saving and retrieving files, using a web browser and an email program. Today, it is rare to encounter students without prior basic experience with computers, and most students already have a computer of their own before entering the study programme.

5. Due to the gender distribution among the participants, some groups consisted of women only, while some were mixed.

6. The analysis is based on my doctoral thesis. The presentation is, however, restructured along the described continuum of harmony–disharmony–rejection. Quotes from emails and interview transcripts have been translated from Norwegian and they have been linguistically normalized and repetitive words and sounds have been left out. Unlike in my doctoral thesis, in this presentation I have chosen not to emphasize the individuals, but rather to focus on the kind of rhetoric used in the various positioning strategies. For a more nuanced description of this material and more expanded analysis, see the PhD study: Corneliussen (2003a – in Norwegian), or Corneliussen 2003c –in English) with an expanded analysis of the male students.

7. Cf. Lagesen (2005), who discusses gendered perceptions of computing in Malaysia, which, different from Western perceptions, sees computing as a suitable study for women.

5   Stories about Individual Change and Transformation

1. In this chapter the informants have been given fictitious names to be able to talk about them individually.

2. At the time of this study female students comprised between 60 and 70 per cent of those attending undergraduate courses at the Department of Humanistic Informatics.

3. Håpnes and Rasmussen have reported a similar incident from an interview with two schoolgirls, where one of them says that she enjoys playing computer games. The other girl is more sceptical and refers to games as a thing for boys. The first girl immediately moderates her story, claiming that she does not play that often (as boys do), and not for that long, and not alone (Håpnes and Rasmussen, 2003). This illustrates that girls might have fun with computers, but in this case it was out of line with the unwritten rules for being a young girl.

6 Layered Meanings and Differences Within

1. West and Zimmerman's starting point is what they found to be a problematic schism of the 1960s and early 1970s between sex and gender, where the first was 'ascribed by biology' and the other 'an achieved status' (1987, p. 125). This confused their students, they claim, because sex seemed less a 'given' and gender seemed less an 'achievement', and 'the relationship between biological and cultural processes was far more complex and reflexive than we previously had supposed' (1987, pp. 125–6).

2. Of the Norwegian population, 77 per cent had access to a computer at home in 2003 (see Chapter 2). In 2008, 91 per cent of men and 90 per cent of women had access to a computer at home (SSB).

3. Other differentiating features are also introduced in a number of cases (such as age, education and so on), but they are not repeated as persistently as gender, and they do not really enter the discursive universe of taken-for-granted assumptions about computer users.

4. This slogan is from the recruitment campaign at NTNU in 1997, which used this as the final slide in a cinema commercial (cf. Chapter 5; Lagesen, 2003).

5. See, for example, the comments to the articles '38% of gamers are women' (http://www.computerandvideogames.com/article.php?id=183778, accessed June 2010) and 'Study: 64% of Online Gamers are Women' (http://www.1up.com/do/newsStory?cId=3154239, accessed June 2010).

7 Is There an Elsewhere?

1. Internet World Stats, http://www.internetworldstats.com/, accessed December, 2010.

2. According to Sysomos, women make up 50.9 per cent of the blogosphere (2010). Out of the 20 most accessed blogs in the Norwegian blogging community blogg.no, the first 15 are created by girls or young women (http://blogg.no/#toplist-visited, accessed 27 November 2010). Blog number 1 on the toplist for blogg.no is written by a girl aged 15. Her blog has been the most popular Norwegian blog for some considerable time, with more than 50,000 hits per day, and she has appeared in national newspapers several times since she first started blogging in 2009 (*Dagbladet* 5 April, 24 May

2010, *VG* 15 June 2010). Figures for 2009 show that women also outnumber men on Facebook in the US at 56.2 per cent, and with women aged over 55 as the fastest growing group of new Facebook users (Inside Facebook, February 2009 http://www.insidefacebook.com/2009/02/02/fastest-growing-demographic-on-facebook-women-over-55/, accessed 12 December 2010). Men outnumber women on Twitter, however, with women following close behind, at 47 per cent, and among the top 5 per cent, men dominate at 54 per cent against 46 per cent for women (Sysomos, 2009).

3. The critical mass is not the same as the mass you need for something to *be* gender equal, but the mass needed to not make women stand out as gendered targets or icons, making them more exposed to ruling techniques meant to put women out of play by emphasizing their gender instead of their skills (Powell et al., 2006). Thus a critical mass can be seen as a proportion of women sufficiently large for women to feel comfortable.

4. NSD figures for 2002–09 show that the proportion of female students slowly and steadily increased from 59.6 per cent in 2002 to 60.9 per cent in 2009 (reaching a high of 61.3 per cent in 2008). The proportion of women in the category 'Natural sciences, crafts and technical subjects' increased from 29.7 per cent in 2002 to 32.4 per cent in 2009. The number of women in 'information and computer technology' has remained more or less stable at 21.4 per cent in 2002, reaching a high of 23.3 per cent in 2008, and dropping back to 21.3 per cent in 2009. In 'electromechanical, machine subjects' the number of women has increased from 17.3 per cent in 2002 to 20.3 per cent in 2009.

5. See, for instance, the article 'The most feared woman on the Internet', published at Salon.com in 2002, which questions the reality in stories about Netochka Nezvanova. An editor's note has been added to the article since the first time I visited it in 2002: 'After the publication of this story, Salon received a lengthy e-mail from someone presenting herself as Netochka Nezvanova, challenging much of the article. ... We have corrected several facts that we were able to corroborate .... But as proved true during the original reporting of the story, and as we tried to communicate in the article itself, there's a level of obscurity in the communications of Ms. Nezvanova that makes ascertaining the truth a challenge.' http://www.salon.com/technology/feature/2002/03/01/netochka/index.html, accessed June 2010.

6. http://kifinfo.no/nyhet/vis.html?tid=72272, accessed June 2010.

7. 'Inappropriate/d Artificiality', interview with Trinh T. Minh-ha by Marina Grzinic, http://arch.ced.berkeley.edu/people/faculty/bourdier/trinh/TTMHInterviews002.htm, accessed July 2010.

8. Ibid.

References

AUWW (2000) *Tech-Savvy: Educating Girls in the New Computer Age* (Washington, DC: American Association of University Women (AAUW) Educational Foundation).

Abbate, J. (2003) 'Guest Editor's Introduction: Women and Gender in the History of Computing'. *IEEE Annals of the History of Computing*, 25, 4, 4–8.

Abbate, J. (2010) 'The Pleasure Paradox: Bridging the Gap between Popular Images of Computing and Women's Historical Experiences'. In Misa, T. J. (ed.) *Gender Codes: Why Women are Leaving Computing* (Hoboken, NJ: IEEE Computer Society and John Wiley & Sons, Inc.): 213–28.

Adam, A., Emms, J., Green, E. and Owen, J. (1994) *Women, Work and Computerization. Breaking Old Boundaries, Building New Forms.* Proceedings of the IFIP TC9/WG9.1 Fifth International Conference on Women, Work and Computerization, Manchester, U.K., 2–5 July, 1994 (Amsterdam: North-Holland).

Adam, A., Griffiths, M., Keogh, C., Moore, K., Richardson, H. and Tattersall, A. (2005) ' "You Don't Have To Be Male To Work Here, But It Helps!" – Gender and the IT Labour Market'. In Archibald, J., Emms, J., Grundy, F., Payne, J. and Turner, E. (eds) *The Gender Politics of ICT* (Middlesex: Middlesex University Press): 283–96.

Adam, A., Howcroft, D. and Richardson, H. (2004) 'A Decade of Neglect: Reflecting on Gender and IS'. *New Technology, Work and Employment*, 19, 3, 222–40.

Ahuja, M. (2002) 'Women in the Information Technology Profession: A Literature Review, Synthesis and Research Agenda'. *European Journal of Information Systems*, 11, 20–34.

Akrich, M. (1992) 'The De-Scription of Technical Objects'. In Bijker, W. E. and Law, J. (eds) *Shaping Technology/Building Society. Studies in Sociotechnical Change* (Cambridge, MA: MIT Press): 205–24.

Akrich, M. and Latour, B. (1992) 'A Summary of a Convenient Vocabulary for the Semiotics of Human and Nonhuman Assemblies'. In Bijker, W. and Law, J. (eds) *Shaping Technology, Building Society: Studies in Sociotechnical Change* (Cambridge, MA: MIT Press): 259–64.

Alpay, E., Ahearn, A. L., Graham, R. H. and Bull, A. M. J. (2008) 'Student Enthusiasm for Engineering: Charting Changes in Student Aspirations and Motivation'. *European Journal of Engineering Education*, 33, 5–6, 573–85.

Annfelt, T. (1999) *Kjønn i utdanning. Hegemoniske posisjoner og forhandlinger om yrkesidentitet i medisin- og faglærerutdanning* (Trondheim: NTNU, Senter for kvinneforskning 2/99).

Asdal, K., Berg, A.-J., Brenna, B., Moser, I. and Rustad, L. (1998) *Betatt av viten – bruksanvisninger til Donna Haraway* (Oslo: Spartacus Forlag A/S).

Asdal, K. and Myklebust, S. (1999) *Teknologi, vitenskap og makt: eksempler på effektiv "black boxing" og noen åpningsforsøk. Et posisjonspapir for Makt- og demokratiutredningen* (Oslo:, Makt- og demokratiutredningen 1998–2003).

Ashcraft, C. and Blithe, S. (2010) 'Women in IT: The Facts' (National Center for Women & Information Technology www.ncwit.org).

Aune, M. (1996) 'The Computer in Everyday Life. Patterns of Domestication of a New Technology'. In Lie, M. and Sørensen, K. H. (eds) *Making technology our own? Domesticating technology into everyday life* (Oslo: Scandinavian University Press): 91–120.

Austgulen, R. (1996) 'Damer @ data – tid for spesielle tiltak?'. *Universitetsavisa, NTNU*, 6, 17 http://www.ntnu.no/universitetsavisa/nr17/innspill.html (accessed 13 February 2003).

Bagilhole, B., Barnard, S., Dainty, A. and Hassan, T. (2010) 'Engineering Curriculum in UK Higher Education and Gender: A Quantitative and Qualitative Exploration of the Effects of Interdisciplinary Content'. In Proceedings of *Beyond the Leaky Pipeline. Challenges for Research on Gender and Science*, Final conference of the study 'Meta-analysis of gender and science research', 19–20 October 2010, Brussels (Belgium: unpublished).

Bagilhole, B., Powell, A., Barnard, S. and Dainty, A. (2008) 'Researching Cultures in Science, Engineering and Technology: An Analysis of Current and Past Literature'. *UKRC Research Report Series, No. 7* (UK: Resource Centre for Women in Science, Engineering and Technology (UKRC)).

Bakhtin, M. M. (1986) 'The Problem of Speech Genres'. In Emerson, C. and Holquist, M. (eds) *Speech Genres and Other Late Essays* (Austin, TX: University of Texas Press.

Barad, K. (2007) *Meeting the Universe Halfway. Quantum Physics and the Entanglement of Matter and Meaning* (Durham and London: Duke University Press).

Barker, J. and Downing, H. (1985 [1980]) 'Word Processing and the Transformation of Patriarchal Relations of Control in the Office'. In MacKenzie, D. and Wajcman, J. (eds) *The Social Shaping of Technology* (Milton Keynes: Open University Press): 147–64.

Barker, L. J. and Aspray, W. (2006) 'The State of Research on Girls and IT'. In Cohoon, J. M. and Aspray, W. (eds) *Women and Information Technology. Research on Underrepresentation* (Cambridge, MA and London: MIT Press): 3–54.

Baudrillard, J. (1994) *Simulacra and Simulation* (Ann Arbor, MI: University of Michigan Press).

Bauman, Z. (2000) *Liquid Modernity* (Cambridge: Polity Press).

Beauvoir, S. de (1989 [1949]) *The Second Sex* (New York: Vintage Books).

Beck, U. (1992) *Risk Society: Towards a New Modernity* (London: Sage).

Becta (2008) How Do Boys and Girls Differ in their Use of ICT? (Research report http://becta.org.uk).

Bell, D. (2007) *Cyberculture Theorists. Manuel Castells and Donna Haraway* (London and New York: Routledge).

Berg, A.-J. (1996) *Digital Feminism* (Trondheim: NTNU, Institutt for sosiologi og statsvitenskap 28/96).

Berg, A.-J. (1999) 'A Gendered Socio-Technical Construction: The Smart House'. In MacKenzie, D. and Wajcman, J. (eds) *The Social Shaping of Technology* 2nd edn (Buckingham, PA: Open University Press): 301–13.

Berker, T., Hartmann, M., Punie, Y. and Ward, K. (eds) (2006) *Domestication of Media and Technology* (Maidenhead: Open University Press).

Berner, B. and Mellström, U. (1997) 'Looking for Mister Engineer. Understanding Masculinity and Technology at Two Fin de Siècles'. In Berner, B. (ed.) *Gender Practices. Feminist Studies of Technology and Society* (Linköping University, Almqvist & Wiksell International): 39–68.

Bijker, W. and Law, J. (eds) (1992) *Shaping Technology/Building Society. Studies in Sociotechnical Change* (Cambridge, MA: MIT Press).

Bijker, W. E., Huges, T. P. and Pinch, T. (eds) (1993 [1987]) *The Social Construction of Technological Systems* (Cambridge, MA: MIT Press).

Björkman, C. (2002) *Challenging Canon: The Gender Question in Computer Science* (Karlskrona: Department of Human Work Science and Media Technology Blekinge Institute of Technology).

Björkman, C. (2005) 'Invitation to Dialogue: Feminist Research Meets Computer Science'. In Archibald, J., Emms, J., Grundy, F., Payne, J. and Turner, E. (eds) *The Gender Politics of ICT* (Middlesex: Middlesex University Press): 223–36.

Björkman, C., Christoff, I., Palm, F. and Vallin, A. (1997) 'Exploring the Pipeline. Towards an Understanding of the Male Dominated Computing Culture and its Influence on Women'. In Lander, R. and Adam, A. (eds) *Women in Computing* (Exeter, UK: intellect™): 50–9.

Braa, K. (1996) 'Jentene flykter fra informatikk'. *Dagbladet*, http://www.dagbladet.no/kronikker/961029-kro-1.html (accessed 8 February 2003).

Braidotti, R. (1994) *Nomadic Subjects: Embodiment and Sexual Difference in Contemporary Feminist Theory* (New York, Columbia University Press).

Braidotti, R. (2002) *Metamorphoses: Towards a Materialist Theory of Becoming* (Cambridge: Polity Press).

Braidotti, R. (2006) *Transpositions: On Nomadic Ethics* (Malden, MA: Polity Press).

Bratteteig, T. (2002) 'Bringing Gender Issues to Technology Design'. In Floyd, C., Kelker, G., Klein-Franke, Kramarae, C. and Limpangog, C. (eds) *Feminist Challenges in the Information Age* (Opladen: Leske & Budrich): 91–105.

Bratteteig, T. and Verne, G. (1985) 'Kvinneperspektiv på informatikk – teori vs. praksis?'. *Nytt om kvinneforskning*, 5, 9–17.

Bratteteig, T. and Verne, G. (1997) 'Feministisk eller bare kritisk?'. *Kvinneforskning*, 17, 2, 11–24.

British Computer Society (BCS) e-skills UK and Intellect (2009) Women in IT Scorecard. (http://www.intellectuk.org/content/view/4956/47/ (accessed 21 November 2010)).

Bruns, A. (2008) *Blogs, Wikipedia, Second Life, and Beyond: From Production to Produsage* (New York: Peter Lang).

Butler, J. (1990) *Gender Trouble: Feminism and the Subversion of Identity* (New York: Routledge).

Butler, J. (2004) *Undoing Gender* (New York: Routledge).

Camp, T. (1997) 'The Incredible Shrinking Pipeline'. *Communications of the ACM*, 40, 10, 103–10.

Camp, T. and Gürer, D. (2002) Investigating the Incredible Shrinking Pipeline for Women in Computer Science. (Final Report, NSF Project 9812016).

Campbell-Kelly, M. and Aspray, W. (2004) *Computer: A History of the Information Machine* (Boulder, CO: Westview Press).

Carter, R. and Kirkup, G. (1990) *Women in Engineering: A Good Place To Be?* (Basingstoke: Palgrave Macmillan).

Cassell, J. and Jenkins, H. (1998) 'Chess for Girls? Feminism and Computer Games'. In Cassell, J. and Jenkins, H. (eds) *From Barbie to Mortal Kombat. Gender and Computer Games* (Cambridge, MA: MIT Press): 2–45.

Castells, M. (1996) *The Rise of the Network Society* (Oxford: Blackwell).

Castells, M. (2001) *The Internet Galaxy. Reflections on the Internet, Business, and Society* (Oxford: Oxford University Press).

Ceruzzi, P. E. (1998) *A History of Modern Computing* (Cambridge, MA: MIT Press).

Ceruzzi, P. E. (1999) 'Inventing Personal Computing'. In MacKenzie, D. and Wajcman, J. (eds) *The Social Shaping of Technology*, 2nd edn (Buckingham, PA: Open University Press): 64–86.

Charles, M. and Bradley, K. (2006) 'A Matter of Degrees: Female Underrepresentation in Computer Science Programs Cross-Nationally'. In Cohoon, J. M. and Aspray, W. (eds) *Women and Information Technology. Research on Underrepresentation* (Cambridge, MA and London: MIT Press): 183–203.

Cheryan, S., Plaut, V. C., Davies, P. G. and Steele, C. M. (2009) 'Ambient Belonging: How Stereotypical Cues Impact Gender Participation in Computer Science'. *Journal of Personality and Social Psychology*, 97, 6, 1045–60.

Cockburn, C. (1983) *Brothers: Male Dominance and Technological Change* (London: Pluto Press).

Cockburn, C. (1985) *Machinery of Dominance. Women, Men, and Technical Know-How* (London: Pluto Press).

Cockburn, C. (1988 [1985]) *Machinery of Dominance. Women, Men, and Technical Know-How* (Boston: Northeastern University Press).

Cockburn, C. (1992) 'The Circuit of Technology: Gender, Identity and Power'. In Silverstone, R. and Hirsch, E. (eds) *Consuming Technologies: Media and Information in Domestic Spaces* (London: Routledge): 32–47.

Cockburn, C. (1999 [1983]) 'Caught in the Wheels. The High Cost of being a Female Cog in the Male Machinery of Engineering'. In MacKenzie, D. and Wajcman, J. (eds) *The Social Shaping of Technology*. 2nd edn (Maidenhead: Open University Press): 126–33.

Cockburn, C. and Dilic, R. F. (eds) (1994a) *Bringing Technology Home. Gender and Technology in a Changing Europe* (Buckingham: Open University Press).

Cockburn, C. and Dilic, R. F. (1994b) 'Introduction: Looking for the Gender/ Technology Relation'. In Cockburn, C. and Dilic, R. F. (eds) *Bringing Technology Home. Gender and Technology in a Changing Europe* (Buckingham: Open University Press): 1–21.

Cockburn, C. and Ormrod, S. (1993) *Gender and Technology in the Making* (London and Thousand Oaks, CA: SAGE Publications).

Cohoon, J. M. (2006) 'Just Get Over It or Just Get On with It: Retaining Women in Undergraduate Computing'. In Cohoon, J. M. and Aspray, W. (eds) *Women and Information Technology. Research on Underrepresentation* (Cambridge, Massachusetts, London: MIT Press): 205–37.

Cohoon, J. M. and Aspray, W. (2006a) 'A Critical Review of the Research on Women's Participation in Postsecondary Computing Education'. In Cohoon, J. M. and Aspray, W. (eds) *Women and Information Technology. Research on Underrepresentation* (Cambridge, MA and London: MIT Press): 137–80.

Cohoon, J. M. and Aspray, W. (eds) (2006b) *Women and Information Technology. Research on Underrepresentation* (Cambridge, MA and London: MIT Press).

Connell, R. W. (2002) *Gender* (Cambridge: Polity).

Corneliussen, H. (2002) 'The Multi-Dimensional Stories of the Gendered Users of ICT'. In Morrison, A. (ed.) *Researching ICTs in Context* (Oslo: InterMedia Report): 161–84.

Corneliussen, H. (2003a) *Diskursens makt – individets frihet: Kjønnede posisjoner i diskursen om data (The power of discourse – the freedom of individuals: Gendered positions in the discourse of computing)* (Doctoral Thesis, Department of Humanistic Informatics, University of Bergen).

Corneliussen, H. (2003b) 'Konstruksjoner av kjønn ved høyere IKT-utdanning i Norge'. *Kvinneforskning*, 27, 3, 31–50.

Corneliussen, H. (2003c) 'Male Positioning Strategies in Relation to Computing'. In Lie, M. (ed.) *He, She and IT Revisited. New Perspectives on Gender in the Information Society* (Oslo: Gyldendal Akademisk): 103–34.

Corneliussen, H. (2006) 'Gender in Norwegian Computer History'. In Trauth, E. M. (ed.) *Encyclopedia of Gender and Information Technology* (Hershey, PA: Idea Group Reference): 630–5.

Corneliussen, H. G. (2009) D31 – Country Report: Norway (Meta-analysis of gender and science research, RTD-PP-L4–2007–1).

Cox, C. (1992) 'Eco-Feminism'. In Kirkup, G. and Keller, L. S. (eds) *Inventing Women: Science, Technology and Gender* (Cambridge: Polity Press): 282–93.

Creager, A. N. H., Lunbeck, E. and Schiebinger, L. (2001) 'Introduction'. In Creager, A. N. H., Lunbeck, E. and Schiebinger, L. (eds) *Feminism in Twentieth-Century Science, Technology, and Medicine* (Chicago and London: University of Chicago Press): 1–19.

Crutzen, C. K. M. (1994) 'The Influence of Feminist Theory on Informatics Course Design'. In Adam, A., Emms, J., Green, E. and Owen, J. (eds) *Women, Work and Computerization: Breaking Old Boundaries, Building New Forms* (Amsterdam: Elsevier): 59–73.

Crutzen, C. K. M. (2009) 'The Disappearance of the (Human) Subject in Computer Science'. *Feminist Research Methods – An International Conference* (4–6 February 2009, Stockholm, Sweden).

Denning, P. and McGettrick, A. (2006) 'Re-centering Computer Science – Position Paper for ICER06 '.

Derrida, J. (1978) *Writing and Difference* (London: Routledge & Kegan Paul).

Dovey, J. and Kennedy, H. W. (2006) *Game Cultures. Computer Games as New Media* (Maidenhead: Open University Press).

Dovey, J. and Kennedy, H. W. (2007) 'Technicity: Power and Difference in Game Cultures'. *Games in Action* (Gothenburg, Sweden, 14 June 2007, http://www. dcrc.org.uk/publications/technicity-power-and-difference-game-cultures (accessed 14 October 2010)).

Edelman, B. (1997) 'These Girls are No Ordinary Girls!'. In Berner, B. (ed.) *Gender Practices. Feminist Studies of Technology and Society* (Stockholm: Almqvist & Wiksell International): 19–38.

Egeland, C. (1999) 'Problemet som ikke (vil) finnes'. *Kvinneforskning*, 23, 1, 80–8.

Ellul, J. (1964) *The Technological Society* (New York: Vintage).

Elovaara, P. (2004) *Angels in Unstable Sociomaterial Relations: Stories of Information Technology* (Karlskrona, Blekinge Institute of Technology).

Ensmenger, N. (2010) 'Making Programming Masculine'. In Misa, T. J. (ed.) *Gender Codes: Why Women are Leaving Computing* (Hoboken, NJ: IEEE Computer Society and John Wiley & Sons, Inc.): 115–41.

European Commission (2009) 'SHE figures 2009. Statistics and Indicators on Gender Equality in Science' (Luxembourg: Publications Office of the European Union).

Faulkner, W. (2000a) 'Dualisms, Hierarchies and Gender in Engineering'. *Social Studies of Science*, 30, 5, 759–92.

Faulkner, W. (2000b) 'The Power and Pleasure? A Research Agenda for "Making Gender Stick" to Engineers'. *Science, Technology, & Human Values*, 25, 1, 87–119.

Faulkner, W. (2001) 'The Technology Question in Feminism: A View from Feminist Technology Studies'. *Women's Studies International Forum*, 24, 1, 79–95.

Faulkner, W. (2005) 'Becoming and Belonging: Gendered Processes in Engineering'. In Archibald, J., Emms, J., Grundy, F., Payne, J. and Turner, E. (eds) *The Gender Politics of ICT* (Middlesex: Middlesex University Press): 15–25.

Faulkner, W. (2006) 'Genders in/of Engineering. A research report' (http://www.goodfood-project.org/www/Gender/Faulkner_Genders_in_Engineering_Report.pdf (accessed 8 November 2010)).

Faulkner, W. and Lie, M. (2007) 'Gender in the Information Society: Strategies of Inclusion'. *Gender, Technology and Development*, 11, 2, 157–77.

Firestone, S. (1970) *The Dialectic of Sex: The Case for Feminist Revolution* (New York: Morrow).

Foucault, M. (1999 [1971]) *Diskursens orden* (Oslo: Spartacus Forlag A/S).

Gannon, S. (2007) 'Laptops and Lipsticks: Feminising Technology'. *Learning, Media and Technology*, 32, 1, 53–67.

Gansmo, H. J. (1998) *Det forvrengte dataspeilet* (Trondheim, NTNU, Senter for teknologi og samfunn 36/98).

Gansmo, H. J. (2003) 'Limits of State Feminism. Chaotic Translations of the "Girls and Computing" Problem'. In Lie, M. (ed.) *He, She and IT Revisited. New Perspectives on Gender in the Information Society* (Oslo: Gyldendal Akademisk): 135–72.

Gansmo, H. J. (2004) *Towards a Happy Ending for Girls and Computing?* (Trondheim: Department of Interdisciplinary Studies of Culture, Faculty of Arts, Norwegian University of Science and Technology).

Gansmo, H. J., Lagesen, V. A. and Sørensen, K. H. (2003a) 'Forget the Hacker? A Critical Re-Appraisal of Norwegian Studies of Gender and ICT'. In Lie, M. (ed.) *He, She and IT Revisited. New Perspectives on Gender in the Information Society* (Oslo: Gyldendal Akademisk): 34–68.

Gansmo, H. J., Lagesen, V. A. and Sørensen, K. H. (2003b) 'Out of the Boy's Room? A Critical Analysis of the Understanding of Gender and ICT in Norway'. *NORA*, 11, 3, 130–9.

Gansmo, H. J., Nordli, H. and Sørensen, K. H. (2004) 'The Gender Game. A Study of Norwegian Computer Game Designers'. In Gansmo, H. J. (ed.) *Towards a Happy Ending for Girls and Computing?* (Trondheim, PhD Dissertation).

Gauntlett, D. (2008) *Media, Gender and Identity. An Introduction* (London and New York: Routledge).

Gere, C. (2008) *Digital Culture* (London: Reaktion Books).

Giddens, A. (1984) *The Constitution of Society – Outline of the Theory of Structuration* (Cambridge: Polity Press).

Giddens, A. (1991) *Modernity and Self-Identity. Self and Society in the Late Modern Age* (Cambridge, Polity Press).

Gilbert, S. F. and Rader, K. A. (2001) 'Revisiting Women, Gender, and Feminism in Developmental Biology'. In Creager, A. N. H., Lunbeck, E. and Schiebinger, L. (eds) *Feminism in Twentieth-Century Science, Technology, and Medicine* (Chicago and London: University of Chicago Press): 73–97.

Godfroy-Genin, A.-S. (2009) 'Women's Academic Careers in Technology: A Comparative European Perspective'. *Equal Opportunities International*, 28, 1, 80–97.

Godfroy-Genin, A.-S. and Pinault, C. (2006) 'The Benefits of Comparing Grapefruits and Tangerines: A Toolbox for European Cross-Cultural Comparisons in Engineering Education – Using this Toolbox to Study Gendered Images of Engineering among Students'. *European Journal of Engineering Education*, 31, 1, 23–33.

Grier, D. A. (2005) *When Computers Were Human* (Princeton and Oxford: Princeton University Press).

Grint, K. and Gill, R. (1995) 'The Gender-Technology Relation: Contemporary Theory and Research'. In Grint, K. and Gill, R. (eds) *The Gender-Technology Relation* (London: Taylor & Francis): 1–28.

Grint, K. and Woolgar, S. (1995) 'On Some Failures of Nerve in Constructivist and Feminist Analyses of Technology'. In Grint, K. and Gill, R. (eds) *The Gender-Technology Relation* (London: Taylor & Francis): 48–75.

Grosz, E. (2010) 'The Untimeliness of Feminist Theory'. *NORA*, 18, 1, 48–51.

Grundy, F. (1996) *Women and Computers* (Exeter, UK: intellect™).

Grundy, F. (1997) 'Where Do We Go From Here?'. In Lander, R. and Adam, A. (eds) *Women in Computing* (Exeter, UK: intellect™): 1–10.

Gürer, D. (2002) 'Women in Computing History'. *inroads – SIGCSE Bulletin*, 34, 2, Special Issue Women and Computing, 116–20.

Gürer, D. W. (1995) 'Pioneering Women in Computer Science'. *Communications of the ACM*, 38, 1, 45–54.

Haavind, H. (1986) 'Kvinner i naturvitenskapelig forskning – banebrytere eller gratispassasjerer?'. *Nytt om kvinneforskning*, 1986, 4, 4–12.

Haavind, H. (1993) 'Analyse av kvinners historier – Bearbeiding av makt og splittelse'. In Nielsen, A. M., Haavind, H. and Holter, Ø. G. (eds) *Køn i forandring: ny forskning om køn, socialisering og identitet* (København: Hyldespjæt): 12–44.

Haavind, H. (1994) 'Kjønn i forandring – som fenomen og som forståelsesmåte'. *Tidsskrift for Norsk Psykologiforening*, 31, 767–83.

Haavind, H. (2000) 'Analytiske retningslinjer ved empiriske studier av kjønnede betydninger'. In Haavind, H. (ed.) *Kjønn og fortolkende metode. Metodiske muligheter i kvalitativ forskning* (Oslo: Gyldendal Norsk Forlag AS): 155–219.

Hacker, S. (1989) *Pleasure, Power, and Technology. Some Tales of Gender, Engineering, and the Cooperative Workplace* (Boston: Unwin Hyman).

Haddon, L. (1991) Researching Gender and Home Computers. In Sørensen, K. H. and Berg, A.-J. (eds) *Technology and Everyday Life: Trajectories and Transformations* (Report No. 5: Proceedings from a Workshop in Trondheim 1990).

Haigh, T. (2010) 'Masculinity and the Machine Man: Gender in the History of Data Processing'. In Misa, T. J. (ed.) *Gender Codes: Why Women are Leaving Computing* (Hoboken, NJ: IEEE Computer Society and John Wiley & Sons, Inc.): 51–71.

Haraway, D. (1991a) 'A Cyborg Manifesto'. In Haraway, D. (ed.) *Simians, Cyborgs, and Women. The Reinvention of Nature* (London: Free Association Books): 149–81.

Haraway, D. (1991b) *Simians, Cyborgs, and Women. The Reinvention of Nature* (London: Free Association Books).

Haraway, D. (1991c) 'Situated Knowledges. The Science Question in Femininsm and the Privilege of Partial Perspective'. In Haraway, D. (ed.) *Simians, Cyborgs, and Women. The Reinvention of Nature* (London: Free Association Books): 183–201.

Haraway, D. (1992) 'The Promises of Monsters: A Regenerative Politics for Inappropriate/d Others'. In Grossberg, L., Nelson, C. and Treichler, P. (eds) *Cultural Studies* (New York: Routledge): 295–337.

Haraway, D. (1997) *Modest_Witness@Second_Millennium.FemaleMan(C)_Meets_Onco Mouse™. Feminism and Technoscience* (New York and London: Routledge).

Haraway, D. (2004) *The Haraway Reader* (New York and London: Routledge).

Haraway, D. J. (2008) *When Species Meet* (Minneapolis and London: University of Minnesota Press).

Harding, S. (1986) *The Science Question in Feminism* (Ithaca, NY: Cornell University Press).

Hausmann, R., Tyson, L. D. and Zahidi, S. (2010) *The Global Gender Gap Report 2010* (Geneva, Switzerland: World Economic Forum).

Hayes, C. C. (2010a) 'Computer Science: The Incredible Shrinking Woman'. In Misa, T. J. (ed.) *Gender Codes: Why Women are Leaving Computing* (Hoboken, NJ: IEEE Computer Society and John Wiley & Sons, Inc.): 25–49.

Hayes, C. C. (2010b) 'Gender Codes: Prospects for Change'. In Misa, T. J. (ed.) *Gender Codes: Why Women are Leaving Computing* (Hoboken, NJ: IEEE Computer Society and John Wiley & Sons, Inc.): 265–73.

Hernes, H. M. (1987) *Welfare State and Woman Power. Essays in State Feminism* (Oslo: Norwegian University Press).

Hofstede, G. (2003 [1991]) *Cultures and Organizations, Software of the Mind: Intercultural Cooperation and its Importance for Survival* (London: Profile Books).

Holter, Ø. G. (2010) 'Gender Equality and the Modernisation of Academic Organisations'. In Rustad, L. M. and Winsnes Rødland, A. (eds) *Talent at Stake. Changing the Culture of Research – Gender-Sensitive Leadership* (Committee for Gender Balance in Research http://eng.kifinfo.no (accessed 12 November 2010)).

Holter, Ø. G., Svare, H. and Egeland., C. (2009) 'Gender Equality and Quality of Life – A Norwegian Perspective'. Nordisk Institutt for kunnskap om kjønn (NIKK).

Hughes, T. P. (1986) 'The Seamless Web: Technology, Science, Etcetera, Etcetera'. *Social Studies of Science,* 16, 2, 281–92.

Huws, U. (1991) 'Telework: Projections'. *Futures,* 23, 1, 19–31.

Håpnes, T. (1992) 'Hvordan forstå mannsdominansen i datafaget? En dekonstruksjon av fag- og kjønnskultur'. In Annfelt, T. and Imsen, G. (eds) *Utdanningskultur og kjønn* (Trondheim, NTNU, Senter for teknologi og samfunn 3/92): 155–83.

Håpnes, T. (1996) 'Not in Their Machines. How Hackers Transform Computers into Subcultural Artefacts'. In Lie, M. and Sørensen, K. H. (eds) *Making Technology Our Own? Domesticating Technology into Everyday Life* (Oslo: Scandinavian University Press): 121–50.

Håpnes, T. (1997) Kommunikasjonsteknologi og likestilling (Report, NORUT Samfunnsforskning).

Håpnes, T. and Rasmussen, B. (1990) Har datafaget kjønn? (Trondheim: Institutt for industriell miljøforskning Notat 6/90).

Håpnes, T. and Rasmussen, B. (1999) 'Jenteidentitet på internett'. *Sosiologisk Tidsskrift,* 7, 1, 3–21.

Håpnes, T. and Rasmussen, B. (2003) 'Gendering Technology. Young Girls Negotiating ICT and Gender'. In Lie, M. (ed.) *He, She and IT Revisited. New Perspectives on Gender in the Information Society* (Oslo: Gyldendal Akademisk): 173–97.

Jenkins, H. (2006) *Convergent Culture – Where Old and New Media Collide* (New York and London: New York University Press).

Kanter, R. M. (1993 [1977]) *Men and Women of the Corporation* (New York: Basic Books).

Kekelis, L. S., Ancheta, R. W. and Heber, E. (2005) 'Hurdles in the Pipeline: Girls and Technology Careers'. *Frontiers: A Journal of Women Studies,* 26, 1, 99–109.

Kirkup, G. and Keller, L. S. (1992) 'The Nature of Science and Technology'. In Kirkup, G. and Keller, L. S. (eds) *Inventing Women: Science, Technology and Gender* (Cambridge: Polity Press): 5–11.

Kitzinger, J., Haran, J., Chimba, M. and Boyce, T. (2008) 'Role Models in the Media: An Exploration of the Views and Experiences of Women in Science, Engineering and Technology' (UK Resource Centre for Women in Science, Engineering and Technology (UKRC) and Cardiff University).

Kleif, T. and Faulkner, W. (2003) ' "I'm No Athlete [but] I Can Make This Thing Dance!" – Men's Pleasures in Technology'. *Science, Technology, & Human Values,* 28, 2, Spring 2003, 296–325.

Kvaløy, K. (1999) *Fortellinger om moderne flinke lekne ungdomsjenters forhold til datate nologi. En kvalitativ studie av datateknologiens rolle i ungdomsjenters dannelse av kjønnsidentitet* (Trondheim: NTNU, Senter for kvinneforskning 3/99).

Kvande, E. (1982) Kvinner og høgere teknisk utdanning. Delrapport: rekruttering og rekrutteringstiltak (Trondheim: Universitetet i Trondheim Norges tekniske høgskole, IFIM Institutt for industriell miljøforskning tilsluttet SINTEF).

Kvande, E. (1984) Kvinner og høgere teknisk utdanning – Integrert eller utdefinert? Om kvinnelige NTH-studenters studiesituasjon og framtidsplaner (Trondheim: IFIM).

Kvande, E. and Rasmussen, B. (1993 [1990]) *Nye kvinneliv. Kvinner i menns organisasjoner* (Oslo: Ad Notam).

Laclau, E. (1990) *New Reflections on the Revolution of Our Time* (London and New York: Verso).

Laclau, E. and Mouffe, C. (1985) *Hegemony and Socialist Strategy. Towards a Radical Democratic Politics* (London: Verso).

Lagesen Berg, V. A. (2000) *Firkanter og rundinger. Kjønnskonstruksjoner blant kvinnelige dataingeniørstudenter ved NTNU* (Trondheim: NTNU, Senter for kvinneforskning 3/00).

Lagesen Berg, V. A. and Kvaløy, K. (1998) En kvantitativ undersøkelse av trivsel og studiemotivasjon blant førsteårsstudentene ved linjen for datateknikk, NTNU, samt en evaluering av fagmodulen "Kjenn ditt fag" (Trondheim: NTNU, Institutt for datateknikk og informasjonsvitenskap).

Lagesen, V. A. (2003) 'Advertising Computer Science to Women (Or Was it the Other Way Around?)'. In Lie, M. (ed.) *He, She and IT Revisited. New Perspectives on Gender in the Information Society* (Oslo: Gyldendal Akademisk): 69–102.

Lagesen, V. A. (2005) *Extreme Make-Over? The Making of Gender and Computer Science* (Trondheim: Department of Interdisciplinary Studies of Culture Faculty of Arts Norwegian University of Science and Technology).

Lagesen, V. A. (2007) 'The Strength of Numbers: Strategies to Include Women into Computer Science'. *Social Studies of Science*, 37, 1, 67–92.

Lagesen, V. A. and Sørensen, K. H. (2009) 'Walking the Line? The Enactment of the Social/Technical Binary in Software Engineering'. *Engineering Studies*, 1, 2, 129–49.

Lander, R. and Adam, A. (eds) (1997) *Women in Computing* (Exeter, UK: intellect™).

Langsether, H. (2001) Behov og barrierer for jenter på informatikkstudiet. *Skriftserie 2/2001* (Trondheim: Senter for kvinne- og kjønnsforskning, NTNU).

Leavitt, D. (2006) *The Man Who Knew Too Much: Alan Turing and the Invention of the Computer* (New York: Norton).

Levold, N. (2001) '"Doing Gender"' in Academia. The Domestication of an Information-Technological Researcher-Position'. In Glimell, H. and Juhlin, O. (eds) *The Social Production of Technology. On Everyday Life with Things* (Gothenburg: BAS Publisher): 133–58.

Levold, N. (2002) 'Mellom lyst og plikt? Beretning om en mannlig IKT-forsker'. *Kvinneforskning*, 2002, 2, 50–65.

Levy, S. (1984) *Hackers: Heroes of the Computer Revolution* (New York: Bantam Doubleday Dell Publishing. Group).

Lie, M. (1996) 'Gender in the Image of Technology'. In Lie, M. and Sørensen, K. H. (eds) *Making Technology Our Own? Domesticating Technology into Everyday Life* (Oslo: Scandinavian University Press): 201–23.

Lie, M. (1998) *Computer Dialogues. Technology, Gender and Change* (Trondheim: NTNU, Senter for kvinneforskning 2/98).

Lie, M. (ed.) (2003a) *He, She and IT revisited. New Perspectives on Gender in the Information Society* (Oslo: Gyldendal Akademisk).

Lie, M. (2003b) 'The New Amazons. Gender Symbolism on the Net'. In Lie, M. (ed.) *He, She and IT Revisited. New Perspectives on Gender in the Information Society* (Oslo: Gyldendal Akademisk): 251–77.

Lie, M., Berg, A.-J., Kaul, H., Kvande, E., Rasmussen, B. and Sørensen, K. H. (1984) 'Har teknologi noe med kvinner å gjøre?'. *Sosiolog i dag*, 1, 23–39.

Lie, M., Berg, A.-J., Kaul, H., Kvande, E., Rasmussen, B. and Sørensen, K. H. (1988) *I menns bilde. Kvinner – teknologi – arbeid* (Trondheim: Tapir).

Lie, M. and Sørensen, K. H. (1996a) 'Making Technology Our Own? Domesticating Technology into Everyday Life'. In Lie, M. and Sørensen, K. H. (eds) *Making Technology Our Own? Domesticating Technology into Everyday Life* (Oslo: Scandinavian University Press): 1–30.

Lie, M. and Sørensen, K. H. (eds) (1996b) *Making Technology Our Own? Domesticating Technology into Everyday Life* (Oslo: Scandinavian University Press).

Light, J. S. (1999) 'When Computers Were Women'. *Technology and Culture*, 40, 3, 455–83.

Lockheed, M. E. (1985) 'Women, Girls, and Computers: A First Look at the Evidence'. *Sex Roles*, 13, 3/4, 115–22.

Lohan, M. (2000) 'Constructive Tensions in Feminist Technology Studies'. *Social Studies of Science*, 30, 6, 895–916.

Lykke, N., Markussen, R. and Olesen, F. (2004) 'Cyborgs, Coyotes and Dogs: A Kinship of Feminist Figurations and There Are Always More Things Going on Than You Thought! Methodologies as Thinking Technologies'. *The Haraway Reader* (London and New York: Routledge): 321–42.

Maass, S., Rommes, E., Schirmer, C. and Zorn, I. (2007) 'Gender Research and IT Construction: Concepts for a Challenging Partnership'. In Zorn, I., Maass, S., Rommes, E., Schirmer, C. and Schelhowe, H. (eds) *Gender Designs IT: Construction and Deconstruction of Information Society Technology* (Wiesbaden: VS Verlag für Sozialwissenschaften): 9–32.

Mack, P. E. (2001) 'What Difference Has Feminism Made to Engineering in the Twentieth Century?'. In Creager, A. N. H., Lunbeck, E. and Schiebinger, L. (eds) *Feminism in Twentieth-Century Science, Technology, and Medicine* (Chicago and London: University of Chicago Press): 149–68.

Mahoney, M. S. (2001) 'Boys' Toys and Women's Work: Feminism Engages Software'. In Creager, A. N. H., Lunbeck, E. and Schiebinger, L. (eds) *Feminism in Twentieth-Century Science, Technology, and Medicine* (Chicago and London: University of Chicago Press): 169–85.

Margolis, J. and Fisher, A. (2002) *Unlocking the Clubhouse. Women in Computing* (Cambridge, MA: MIT Press).

McIlwee, J. S. and Robinson, J. G. (1992) *Women in Engineering: Gender, Power, and Workplace Culture* (Albany: State University of New York Press).

Mellström, U. (1995) *Engineering Lives: Technology, Time and Space in a Male-Centred World* (Linköping: Linköping University).

Mellström, U. (1996) 'Teknologi och maskulinitet. Män och deras maskiner'. In Berner, B. and Sundin, E. (eds) *Från symaskin till cyborg. Genus, teknik och social förändring* (Stockholm: Nerenius and Santérus): 113–39.

Mellström, U. (1999) *Män och deras maskiner* (Nora: Bokförlaget Nya Doxa).

Mellström, U. (2004) 'Machines and Masculine Subjectivity: Technology as an Integral Part of Men's Life Experiences'. *Men and Masculinities*, 6, 4, 368–82.

Meulders, D., Plasman, R., Rigo, A. and O'Dorchai, S. (2010) 'Horizontal and Vertical Segregation' (Meta-analysis of gender and science research, http://www.genderandscience.org/web/reports.php).

Mills, S. (1997) *Discourse* (London: Routledge).

Misa, T. J. (2010a) 'Gender Codes: Lessons from History'. In Misa, T. J. (ed.) *Gender Codes: Why Women are Leaving Computing* (Hoboken, NJ: IEEE Computer Society and John Wiley & Sons, Inc.): 251–63.

190   *References*

Misa, T. J. (ed.) (2010b) *Gender Codes: Why Women are Leaving Computing* (Hoboken, NJ: IEEE Computer Society and John Wiley & Sons, Inc.).

Moi, T. (1999) *What is a Woman? And Other Essays* (Oxford: Oxford University Press).

Mörtberg, C. (1987) Varför har programmeraryrket blivit manligt? (Forskningsrapport: Tekniska högskolan i Luleå, 1987:42).

Mörtberg, C. (1994) 'Computing as Masculine Culture'. In Gunnarsson, E. and Trojer, L. (eds) *Feminist Voices: On Gender, Technology and Ethics* (Luleå: Luleå University of Technology, Centre for Women's Studies).

Mörtberg, C. (1997) *'Det beror på att man är kvinna ...' Gränsvandrerskor formas och formar informationsteknologi* (Institutionen för Arbetsvetenskap, Avdelningen Genus och Teknik, Luleå tekniska Universitet).

Mörtberg, C. (1999) 'Technoscientific Challenges in Feminism'. *NORA*, 7, 1, 47–62.

Mörtberg, C. (2000) *Where Do We Go From Here? Feminist Challenges of Information Technology* (Luleå: Luleå University of Technology, Division of Gender and Technology).

Mörtberg, C. and Elovaara, P. (2010) 'Attaching People and Technology: Between E and Government'. In Booth, S., Goodman, S. and Kirkup, G. (eds) *Gender Isssues in Learning and Working with Information Technology: Social Constructs and Cultural Contexts* (Hershey, NY: Information Science Reference): 83–98.

Negroponte, N. (1995) *Being Digital* (London: Hodder & Stoughton).

Nerheim, G. and Nordvik, H. W. (1986) *Ikke bare maskiner. Historien om IBM i Norge 1935–1985* (Stavanger: Universitetsforlaget).

Niskanen, K. and Florin, C. (eds) (2010) *Föregångarna. Kvinnliga professorer om liv, makt och vetenskap* (Stockholm: SNS Förlag).

Nissen, J. (1993) *Pojkarna vid datorn. Unga entusiaster i datateknikens värld* (Stockholm: Symposion Graduale).

Nissen, J. (1996) 'Det är klart att det är grabbar som håller på med datorer! Men varför är det så?'. In Sundin, E. and Berner, B. (eds) *Från symaskin till cyborg. Genus, teknik och social förändring* (Stockholm: Nerenius & Santerus Förlag): 141–61.

Nordli, H. (1998) *Fra Spice Girls til Cyber Girls? En kvalitativ studie av datafacinerte [sic] jenter i ungdomsskolen* (Trondheim, NTNU, Senter for teknologi og samfunn 35/98).

Nordli, H. (2003) *The Net is Not Enough. Searching for the Female Hacker* (Trondheim, dr.polit.-avhandling, Institutt for sosiologi og statsvitenskap, NTNU).

Nye, D. E. (2004) 'Technological Prediction: A Promethean Problem'. In Sturken, M., Thomas, D. and Ball-Rokeach, S. J. (eds) *Technological Visions. The Hopes and Fears that Shape New Technologies* (Philadelphia: Temple University Press): 159–76.

O'Leary, D. P. (1999) Accessibility of Computer Science: A Reflection for Faculty Members (http://www.cs.umd.edu/~oleary/faculty/ (accessed 7 February 2003)).

Office for National Statistics (2010) Statistical Bulletin: Internet Access 2010. (ONS).

Ogan, C., Robinson, J. C., Ahuja, M. and Herring, S. C. (2006) 'Gender Differences among Students in Computer Science and Applied Information Technology'. In Cohoon, J. M. and Aspray, W. (eds) *Women and Information*

Technology. Research on Underrepresentation (Cambridge, MA and London: MIT Press): 279–300.

Oldenziel, R. (2001) 'Man the Maker, Woman the Consumer: The Consumption Junction Revisited'. In Creager, A. N. H., Lunbeck, E. and Schiebinger, L. (eds) *Feminism in Twentieth-Century Science, Technology, and Medicine* (Chicago and London: University of Chicago Press): 128–48.

Oldenziel, R., Mohun, A. and Lerman, N. E. (eds) (2003) *Gender & Technology: A Reader* (Baltimore: Johns Hopkins University Press).

Ormrod, S. (1994) '"Let's Nuke the Dinner": Discursive Practices of Gender in the Creation of a New Cooking Process'. In Cockburn, C. and Dilic, R. F. (eds) *Bringing Technology Home: Gender and Technology in a Changing Europe* (Buckingham: Open University Press): 42–58.

Oudshoorn, N., Rommes, E. and Stienstra, M. (2004) 'Configuring the User as Everybody: Gender and Design Cultures in Information and Communication Technologies'. *Science, Technology, & Human Values*, 29, 1, 30–63.

Palfrey, J. and Gasser, U. (2008) *Born Digital. Understanding the First Generation of Digital Natives* (New York: Basic Books).

Pedersen, T. B. (1994) '*Vil* Akademia ha flere kvinner?'. *Nytt om kvinneforskning*, 18, 3, 5–20.

Peters, J., Lane, N., Rees, T. and Samuels, G. (2002) 'A Report on Women in Science, Engineering, and Technology – from The Baroness Greenfield CBE to the Secretary of State for Trade and Industry'. (UK http://www.set4women.gov.uk/set4women/research/the_greenfield_rev.htm (accessed 2 February 2003)).

Pfaffenberg, B. (1988) 'The Social Meaning of the Personal Computer: Or, Why the Personal Computer Revolution Was No Revolution'. *Anthropological Quarterly*, 61, 39–47.

Phipps, A. (2007) 'Re-inscribing Gender Binaries: Deconstructing the Dominant Discourse around Women's Equality in Science, Engineering, and Technology'. *Sociological Review*, 55, 4, 768–87.

Pinch, T. and Bijker, W. E. (1997 [1987]) 'The Social Construction of Facts and Artifacts: Or How the Sociology of Science and the Sociology of Technology Might Benefit Each Other'. In Bijker, W. E., Huges, T. P. and Pinch, T. (eds) *The Social Construction of Technological Systems* (Cambridge, MA: MIT Press): 17–50.

Plant, S. (2000) 'On the Matrix. Cyberfeminist Simulations'. In Bell, D. and Kennedy, B. M. (eds) *The Cybercultures Reader* (London: Routledge): 325–36.

Popper, K. R. (1974) 'The Bucket and the Searchlight: Two Theories of Knowledge'. In Popper, K. R. (ed.) *Objective Knowledge* (New York: Oxford University Press).

Powell, A., Bagilhole, B., Dainty, A. and Neale, R. (2004) 'Does the Engineering Culture in UK Higher Education Advance Women's Careers?'. *Equal Opportunities International*, 23, 7/8, 21–38.

Powell, A., Bagilhole, B. M. and Dainty, A. R. J. (2006) 'The Problem of Women's Assimilation into UK Engineering Cultures: Can Critical Mass Work?'. *Equal Opportunities International*, 25, 8, 688–99.

Prensky, M. (2001) 'Digital Natives, Digital Immigrants'. *On the Horizon*, 9, 5.

Rasmussen, B. (1988) 'Datateknologi – en trussel eller nye muligheter for kvinner på kontor?'. In Lie, M., Berg, A.-J., Kaul, H., Kvande, E., Rasmussen, B. and Sørensen, K. H. (eds) *I menns bilde. Kvinner – teknologi – arbeid* (Trondheim, Tapir forlag): 73–87.

Rasmussen, B. and Håpnes, T. (1991) 'Excluding Women from the Technologies of the Future? A Case-Study of the Culture of Computer Science'. *Futures*, December 1991, 1107–19.

Rees, T. (2001) 'Mainstreaming Gender Equality in Science in the European Union: The 'ETAN Report''. *Gender and Education*, 13, 3, 243–60.

Reland, P. (1997) Data og kvinner (Oslo: Report, Ministry of Children and Family Affairs).

Riley, D. (1996) 'Does A Sex Have A History?'. In Scott, J. W. (ed.) *Feminism & History* (Oxford and New York: Oxford University Press): 17–33.

Robinson, J. G. and McIlwee, J. S. (1991) 'Men, Women, and the Culture of Engineering'. *Sociological Quarterly*, 32, 3, 403–21.

Rommes, E. (2007) '"I'm Not Interested in Computers"': Gender-Based Occupational Choices of Adolescents'. *Information, Communication & Society*, 10, 3, 299–319.

Sagebiel, F. (2003) *New Initiatives in Science and Technology and Mathematics Education at the Formal Level: Masculinity Cultures in Engineering Departments in Institutions of Higher Education and Perspectives for Social Change*. Proceedings of the GASAT 11 International Conference (Mauritius).

Sagebiel, F. and Dahmen, J. (2006) 'Masculinities in Organizational Cultures in Engineering Education in Europe: Results of the European Union Project WomEng'. *European Journal of Engineering Education*, 31, 1, 5–14.

Salminen-Karlsson, M. (1997) 'Why Do They Never Talk About the Girls?'. In Lander, R. and Adam, A. (eds) *Women in Computing* (Exeter, UK: intellect™): 160–72.

Salminen-Karlsson, M. (1999) *Bringing Women into Computing Engineering: Curriculum Reform Processes at Two Institutes of Technology* (Linköping Studies in Education and Psychology No 60, Linköping University Department of Education and Psychology).

Schiebinger, L. (ed.) (2008a) *Gendered Innovations in Science and Engineering* (Stanford, CA: Stanford University Press).

Schiebinger, L. (2008b) 'Introduction: Getting More Women into Science and Engineering – Knowledge Issues'. In Schiebinger, L. (ed.) *Gendered Innovations in Science and Engineering* (Stanford, CA: Stanford University Press): 1–21.

Scott, J. W. (1988) *Gender and the Politics of History* (New York: Columbia University Press).

Scott, J. W. (1993) 'Women's History'. In Kauffman, L. S. (ed.) *American Feminist Thought at Century's End* (Cambridge, MA and Oxford: Blackwell): 234–57.

Scott, J. W. (1996) *Only Paradoxes To Offer. French Feminists and the Rights of Man* (Cambridge, MA: Harvard University Press).

Scott, J. W. (2005) *Parité! Sexual Equality and the Crisis of French Universalism* (Chicago and London: University of Chicago Press).

Shirky, C. (2008) *Here Comes Everybody. The Power of Organizing without Organizations* (New York: Penguin Press).

Silverstone, R., Hirsch, E. and Morley, D. (1997 [1992]) 'Information and Communication Technologies and the Moral Economy of the Household'. In Silverstone, R. and Hirsch, E. (eds) *Consuming Technologies: Media and Information in Domestic Spaces* (London and New York: Routledge): 15–31.

Simondon, G. (1980 [1958]) *On the Mode of Existence of Technical Objects* (London: University of Western Ontario).

Simonsen, D. G. (1996) *Kønnets grænser: poststrukturalistiske strategier – historieteoretiske perspektiver* (Københavns Universitet: Center for Kvinde- og Kønsforskning).

Skelton, C. (2005) 'The "Individualized" (Woman) in the Academy: Ulrich Beck, Gender and Power'. *Gender and Education,* 17, 3, 319–32.

Skjeie, H. and Teigen, M. (2003) *Menn imellom. Mannsdominans og likestillingspolitikk* (Oslo: Gyldendal akademisk).

Smelik, A. and Lykke, N. (eds) (2008) *Bits of Life. Feminism at the Intersections of Media, Bioscience, and Technology* (Seattle and London: University of Washington Press).

Stanley, A. (1998 [1983]) 'Women Hold up Two-Thirds of the Sky. Notes for a Revised History of Technology'. In Hopkins, P. D. (ed.) *Sex/machine. Readings in culture, gender, and technology* (Bloomington, IN: Indiana University Press): 17–32.

Statistics Norway (2011) *Statistical Yearbook of Norway 2010* (http://www.ssb.no/english/yearbook/).

Staunæs, D. and Søndergaard, D. M. (2008) 'Management and Gender Diversity. Intertwining Categories and Paradoxes'. In Magnusson, E., Rönnblom, M. and Silius, H. (eds) *Critical Studies of Gender Equalities. Nordic Dislocations, Dilemmas and Contradictions* (Göteborg, Stockholm: Makadam): 135–60.

Steele, C. M. (1997) 'A Threat in the Air. How Stereotypes Shape Intellectual Identity and Performance'. *American Psychologist,* 52, 6, 613–29.

Strathern, M. (1992) 'Foreword: The Mirror of Technology'. In Silverstone, R. and Hirsch, E. (eds) *Consuming Technologies: Media and Information in Domestic Spaces* (London and New York: Routledge): vii–xiii.

Stuedahl, D. (1997a) Jenter og informatikkstudiet – en rapport om jenters studiesituasjon ved Institutt for Informatikk (Universitetet i Oslo).

Stuedahl, D. (1997b) Jenter til informatikkstudiet – utfordringer for utdanningsinstitusjonene. (Presentert på Nokobit-97).

Stuedahl, D. (1999) Studenten i informatikkstudiet – en rapport om studenters situasjon ved Institutt for informatikk, UiO (Oslo: Kirke-, utdannings- og forskningsdepartementet, Likestillingssekretariatet).

Stuedahl, D. and Braa, K. (1997) 'Where Have All The Women Gone – From Computer Science?' (Department of Informatics, University of Oslo).

Sysomos (2009) An In-Depth Look at the 5% of Most Active Users. (http://sysomos.com/insidetwitter/mostactiveusers (accessed 12 December 2010)).

Sysomos (2010) Inside Blog Demographics. (http://sysomos.com/reports/bloggers/ (accessed 12 December 2010)).

Sølvberg, A. (2002) 'Gender Differences in Computer-Related Control Beliefs and Home Computer Use'. *Scandinavian Journal of Educational Research,* 46, 4, 409–26.

Søndergaard, D. M. (1996) *Tegnet på kroppen. Køn: koder og konstruktioner blandt unge voksne i Akademia* (København: Museum Tusculanums Forlag).

Søndergaard, D. M. (1999) *Destabilising discourse analysis: approaches to poststructural empirical research* (København: Institut for Statskundskab Københavns Universitet).

Søndergaard, D. M. (2000) 'Destabiliserende diskursanalyse. Veje ind i poststrukturalistisk inspireret empirisk forskning'. In Haavind, H. (ed.) *Kjønn og fortolkende metode. Metodiske muligheter i kvalitativ forskning* (Oslo, Gyldendal Akademisk): 60–104.

Søndergaard, D. M. (2002) 'Poststructuralist Approaches to Empirical Analysis'. *Qualitative Studies in Education,* 15, 2, 187–204.

Sørensen, K. H. (1991) 'Mot en omsorgspreget teknologi? Om likestillings-politikkens muligheter og begrensninger på et mannsdominert område'. In Haukaa, R. (ed.) *Nye kvinner, nye menn* (Oslo: Ad Notam): 207–27.

Sørensen, K. H. (1992) 'Towards a Feminized Technology? Gendered Values in the Construction of Technology'. *Social Studies of Science,* 22, 5–31.

Sørensen, K. H. (1994) *Technology in Use: Two Essays on the Domestication of Artifacts* (Trondheim: Senter for teknologi og samfunn, Universitetet i Trondheim).

Sørensen, K. H. (2002a) 'Kommunikasjon.no'. *Kvinneforskning,* 2002, 2, 5–9.

Sørensen, K. H. (2002b) Love, Duty and the S-Curve. An Overview of Some Current Literature on Gender and ICT (Strategies of Inclusion: Gender and the Information Society (SIGIS)).

Sørensen, K. H. (2006) 'Domestication: The Enactment of Technology'. In Berker, T., Hartmann, M., Punie, Y. and Ward, K. (eds) *Domestication of Media and Technology* (Maidenhead: Open University Press): 40–61.

Sørensen, K. H. and Lie, M. (1988) 'I menns bilde? Perspektiver på studiet av teknologi og teknikk i kvinners arbeidsliv'. In Lie, M., Berg, A.-J., Kaul, H., Kvande, E., Rasmussen, B. and Sørensen, K. H. (eds) *I menns bilde. Kvinner – teknologi – arbeid* (Trondheim: Tapir forlag): 9–24.

Sørensen, K. H. and Nordli, H. (2005) 'Mobil moral og kjønn i endring? Mobiltelefonen i norske voksnes hverdagsliv'. *Kvinneforskning,* 29, 1.

Teigen, M. (2000) 'Likestilling som legitimeringsstrategi. Rekrutteringsnormer og likestillingspolitikk ved NTNU'. *Sosiologisk tidsskrift,* 2.

Tomas, D. (2000) 'The Technophilic Body: On Technicity in William Gibson's Cyborg Culture'. In Bell, D. and Kennedy, B. M. (eds) *The Cybercultures Reader* (London: Routledge): 175–89.

Torfing, J. (1999) *New Theories of Discourse: Laclau, Mouffe and Žižek* (Oxford: Blackwell).

Trauth, E. M. and Howcroft, D. (2006) 'Critical Empirical Research in IS: An Example of Gender and the IT Workforce'. *Information Technology & People,* 19, 3, 272–92.

Trauth, E. M., Joshi, K. D., Kvasny, L., Chong, J., Kulturel, S. and Mahar, J. (2010) 'Millennials and Masculinity: A Shifting Tide of Gender Typing of ICT'. AMCIS 2010 Proceedings, Paper 73 (http://aisel.aisnet.org/amcis2010/73).

Trauth, E. M., Quesenberry, J. L. and Huang, H. (2009) 'Retaining Women in the U.S. IT Workforce: Theorizing the Influence of Organizational Factors'. *European Journal of Information Systems,* 18, 5, 476–97.

Trinh T. Minh-ha (1989) *Woman, Native, Other: Writing Postcoloniality and Feminism* (Bloomington, IN: Indiana University Press).

Turkle, S. (1984) *The Second Self. Computers and the Human Spirit* (New York: Simon & Schuster).

Turkle, S. (1988) 'Computational Reticence. Why Women Fear the Intimate Machine'. In Kramarae, C. (ed.) *Technology and Women's Voices. Keeping in Touch* (New York: Routledge & Kegan Paul): 41–61.

Turkle, S. (1996) *Life on the Screen. Identity in the Age of the Internet* (London: Weidenfeld & Nicolson).

Turkle, S. (2004) *The Second Self. Computers and the Human Spirit,* 20th anniversary edn (Cambridge, MA and London: MIT Press).

Turkle, S. and Papert, S. (1990) 'Epistemological Pluralism: Styles and Voices Within the Computer Culture' (Massachusetts Institute of Technology 3/90).

Vaage, O. F. (2010) *Norsk mediebarometer 2009* (Oslo, Kongsvinger, Statistisk sentralbyrå – Statistics Norway).

Valenduc, G., Vendramin, P., Guffens, C., Ponzellini, A. M., Lebano, A., D'Ouville, L., Collet, I., Wagner, I., Birbaumer, A., Tolar, M. and Webster, J. (2004) 'Widening Women's Work in Information and Communication Technology'. Synthesis Report (European Commission).

Verne, G. (1988) Rekruttering for enhver pris? *Rekruttering av kvinner til forskning innen informasjonsteknologi og informatikk* (Trondheim: NAVFs sekretariat for kvinneforskning, arbeidsnotat 1/88).

Vestby, G. M. (1998) Jentene, guttene og IT-begrepene. En undersøkelse av ungdoms forståelse av informasjonsteknologi. *NIBR prosjektrapport.* (Oslo).

Virilio, P. (2000) *The Information Bomb* (London: Verso).

Wajcman, J. (1991) *Feminism Confronts Technology* (Cambridge: Polity Press).

Wajcman, J. (2004) *TechnoFeminism* (Cambridge: Polity Press).

Wasburn, M. H. and Miller, S. G. (2006) 'Still a Chilly Climate for Women Students in Technology: A Case Study'. In Fox, M. F., Johnson, D. G. and Rosser, S. V. (eds) *Women, Gender, and Technology* (Urbana and Chicago: University of Illinois Press): 60–79.

Webster, J. (1995) 'What do We Know About Gender and Information Technology at Work? A Discussion of Selected Feminist Research'. *European Journal of Women's Studies*, 2, 3, 315–34.

Webster, J. (1996) *Shaping Women's Work: Gender, Employment and Information Technology* (London: Longman).

Webster, J. (2005) 'Why Are Women Still So Few In IT?'. In Archibald, J., Emms, J., Grundy, F., Payne, J. and Turner, E. (eds) *The Gender Politics of ICT* (Middlesex: Middlesex University Press): 3–14.

Webster, J. (2006) 'Widening of Employment Opportunities in ITEC – Professional Advancement through ITEC Skills' (Report for Equalitec and the Department of Trade and Industry ITEC Skills Team).

Weizenbaum, J. (1976) *Computer Power and Human Reason. From Judgment to Calculation* (San Francisco: Freeman).

West, C. and Zimmerman, D. H. (1987) 'Doing Gender'. *Gender & Society*, 1, 2, 125–51.

White, H. (2003) *Historie og fortelling: utvalgte essay* (Oslo: Pax).

Winther Jørgensen, M. and Phillips, L. (1999) *Diskursanalyse som teori og metode* (Frederiksberg: Roskilde Universitetsforlag Samfundslitteratur).

Woodfield, R. (2000) *Women, Work and Computing* (Cambridge: Cambridge University Press).

Woodfield, R. (2002) 'Women and Information Systems Development: Not Just A Pretty (Inter)Face?'. *Information Technology & People*, 15, 2, 119–138.

Wyatt, S. (2008a) 'Feminism, Technology and the Information Society'. *Information, Communication & Society*, 11, 1, 111–30.

Wyatt, S. (2008b) 'Technological Determinism Is Dead; Long Live Technological Determinism'. In Hackett, E., Amsterdamska, O., Lynch, M. and Wajcman, J. (eds) *The Handbook of Science & Technology Studies* (Cambridge, MA: MIT Press): 165–81.

Åsberg, C. and Lykke, N. (2010) 'Feminist Technoscience Studies'. *European Journal of Women's Studies*, 17, 4, 299–305.

Index